CAMBRIDGE LIBRARY COLLECTION

Books of enduring scholarly value

Life Sciences

Until the nineteenth century, the various subjects now known as the life sciences were regarded either as arcane studies which had little impact on ordinary daily life, or as a genteel hobby for the leisured classes. The increasing academic rigour and systematisation brought to the study of botany, zoology and other disciplines, and their adoption in university curricula, are reflected in the books reissued in this series.

A Naturalist in Western China with Vasculum, Camera and Gun

Ernest Henry Wilson (1876–1930) was introduced to China in 1899 when, as a promising young botanist, he was sent there by horticulturalist Henry Veitch (1840–1924) to collect the seed of the handkerchief tree, *Davidia involucrata*, for propagation in Britain. Subsequent trips saw Wilson bringing back hundreds of seed samples and plant collections, introducing many Chinese plants to Europe and North America. He wrote extensively about his travels in China: this two-volume work was published in 1913. Although much of the text is concerned with plant life, Wilson also gives a great deal of attention to the wider landscape around him. In addition, Wilson took a camera, and these volumes contain photographs of parts of China rarely seen by Europeans in the early twentieth century. In Volume 2 Wilson examines how people in western China use their plants in medicine and agriculture, including the important tea industry.

Cambridge University Press has long been a pioneer in the reissuing of out-of-print titles from its own backlist, producing digital reprints of books that are still sought after by scholars and students but could not be reprinted economically using traditional technology. The Cambridge Library Collection extends this activity to a wider range of books which are still of importance to researchers and professionals, either for the source material they contain, or as landmarks in the history of their academic discipline.

Drawing from the world-renowned collections in the Cambridge University Library, and guided by the advice of experts in each subject area, Cambridge University Press is using state-of-the-art scanning machines in its own Printing House to capture the content of each book selected for inclusion. The files are processed to give a consistently clear, crisp image, and the books finished to the high quality standard for which the Press is recognised around the world. The latest print-on-demand technology ensures that the books will remain available indefinitely, and that orders for single or multiple copies can quickly be supplied.

The Cambridge Library Collection will bring back to life books of enduring scholarly value (including out-of-copyright works originally issued by other publishers) across a wide range of disciplines in the humanities and social sciences and in science and technology.

A Naturalist in Western China with Vasculum, Camera and Gun

Being Some Account of Eleven Years' Travel

VOLUME 2

ERNEST HENRY WILSON

CAMBRIDGE
UNIVERSITY PRESS

CAMBRIDGE UNIVERSITY PRESS

Cambridge, New York, Melbourne, Madrid, Cape Town,
Singapore, São Paolo, Delhi, Tokyo, Mexico City

Published in the United States of America by Cambridge University Press, New York

www.cambridge.org
Information on this title: www.cambridge.org/9781108030465

This edition first published 1913
This digitally printed version 2011

ISBN 978-1-108-03046-5 Paperback

A NATURALIST
IN WESTERN CHINA

THE LOTUS TREE (DIOSPYROS LOTUS) 80 FT. TALL, GIRTH 12 FT.

A NATURALIST IN WESTERN CHINA

WITH VASCULUM, CAMERA, AND GUN

BEING SOME ACCOUNT OF ELEVEN YEARS' TRAVEL,
EXPLORATION, AND OBSERVATION IN THE MORE
REMOTE PARTS OF THE FLOWERY KINGDOM

BY

ERNEST HENRY WILSON, V.M.H.

WITH AN INTRODUCTION BY

CHARLES SPRAGUE SARGENT, LL.D.

WITH ONE HUNDRED AND ONE
FULL-PAGE ILLUSTRATIONS AND A MAP

VOL. II

METHUEN & CO. LTD.
36 ESSEX STREET W.C.
LONDON

First Published in 1913

TO

MY WIFE

CONTENTS

CONTENTS

LIST OF ILLUSTRATIONS

ix

LIST OF ILLUSTRATIONS

A NATURALIST IN WESTERN CHINA

CHAPTER I

THE FLORA OF WESTERN CHINA

A Brief Account of the Richest Temperate Flora in the World

IN previous chapters the wildly mountainous character of Western China has been emphasized. Such a region, affording, as it does, altitudinal extremes, a great diversity in climate, and a copious rainfall, is naturally expected to support a rich and varied flora. Yet after making every allowance for the favourable conditions that obtain in this region the wealth of flowers which meets the eye of the botanist is astonishing and surpasses the dreams of the most sanguine. Competent authorities estimate the Chinese flora to contain fully 15,000 species, half of which are peculiar to the country. These figures speak for themselves and yet fail to give a truly adequate idea of the profusion of flowers. The remote mountain fastnesses of central and Western China are simply a botanical paradise, with trees, shrubs, and herbs massed together in a confusion that is bewildering. On first arriving in a new and strange country it is difficult to recognize the plants one is familiar with under cultivation, and many months necessarily elapse before one is in any sense familiar with the common plants around him. During the eleven years I travelled in China I collected some 65,000 specimens, comprising about 5000 species, and sent home seeds of over 1500 different plants. Nevertheless, it was only during the latter half of this period

that I was able to form an intelligent idea of the flora of China
and to properly appreciate its richness and manifold problems.
The Chinese flora is, beyond question, the richest temperate flora in the world. A greater number of different kinds
of trees are found in China than in the whole of the other
north-temperate regions. Every important genus of broad-
leaved trees known from the temperate regions of the Northern
Hemisphere is represented in China except the Hickory
(*Carya*), Plane (*Platanus*), and False Acacia (*Robinia*). All
the coniferous genera of the same regions, except the Redwoods
(*Sequoia*), Swamp Cypress (*Taxodium*), Chamaecyparis,
Umbrella Pine (*Sciadopitys*), and true Cedars (*Cedrus*), are
found there. In North America, excluding Mexico, about
165 genera of broad-leaved trees occur. In China the number
exceeds 260. Of the 300 genera of shrubs enumerated in
the *Kew Hand-List of Trees and Shrubs* (1902 ed.) fully half
are represented in China.

The great interest and value, however, of the Chinese
flora lies not so much in its wealth of species as in the
ornamental character and suitability of a vast number for
the embellishment of parks and outdoor gardens throughout
the temperate regions of the world. My work in China has
been the means of discovering and introducing numerous new
plants to Europe and North America and elsewhere. But
previous to this work of mine the value of Chinese plants
was well known and appreciated. Evidence of this is afforded
by the fact that there is no garden worthy of the name,
throughout the length and breadth of the temperate parts
of the Northern Hemisphere, that does not contain a few
plants of Chinese origin. Our Tea and Rambler Roses,
Chrysanthemums, Indian Azaleas, Camellias, Greenhouse
Primroses, Moutan Pæonies, and Garden Clematis have all
been derived from plants still to be found in a wild state in
central and Western China. The same is true of a score of
other favourite flowers. China is also the original home of
the Orange, Lemon, Citron, Peach, Apricot, and the so-called
European Walnut. The horticultural world is deeply indebted to the Far East for many of its choicest treasures,
and the debt will increase as the years pass.

DEUTZIA WILSONII

Our knowledge of the marvellous richness of the Chinese flora has been very slowly built up. Travellers, missionaries of all denominations, merchants, consuls, Maritime Customs officials, and all sorts and conditions of men have added their quota; but, as in geography and other departments of knowledge relating to the Far East, the Roman Catholic priests have played the prominent part. The exclusive policy of the Chinese has necessarily increased the difficulties of Europeans who sought to acquire an intimate knowledge of the country, and all honour is due to the workers who have exploited this field in the past.

On behalf of the Royal Horticultural Society of London and others, Robert Fortune, in the 'Forties and 'Fifties of last century, completed the work of his predecessors and exhausted the gardens of China, to our gardens' benefit; but the difficulties of travel were such that he had practically no opportunity of investigating the natural wild flora. With the exception of perhaps half a dozen plants, everything he sent home came from Chinese gardens. But one of his wildlings—*Rhododendron Fortunei*, to wit—has proved of inestimable value to Rhododendron breeders.

Charles Maries, collecting on behalf of Messrs. Veitch, in 1879, ascended the Yangtsze as far as Ichang. He found the natives there unfriendly, and after staying a week was compelled to return. During his brief stay, however, he secured *Primula obconica*, one of the most valuable decorative plants of to-day. Near Kiukiang he secured *Hamamelis mollis*, *Loropetalum chinense*, and a few other plants of less value, and then hied himself away to Japan. For some curious reason or other he concluded that his predecessor, Fortune, had exhausted the floral resources of China, and, most extraordinary of all, his conclusions were accepted! When at Ichang, could he but have gone some three days' journey north, south, or west, he would have secured a haul of new plants such as the botanical and horticultural world had never dreamed of. By the irony of Fate it was left for two or three others to discover and obtain what had been almost within his grasp.

The enormous Chinese population, especially in the vicinity

of the Lower Yangtsze, and its vast alluvial delta and plains, no doubt mizzled Maries, as it has done others. So densely is China populated that every bit of suitable land has been developed under agriculture. A Chinese is capable of getting more returns from a given piece of land than the most expert agriculturist of any other country. Dry farming and intensive cultivation, though unknown to the Chinese under these terms, have been practised by them from time immemorial. The land is never idle, but is always undergoing tilling and manuring. Nevertheless, in spite of the almost incredible industry of the Chinese cultivator, much of the land in the wild mountain fastnesses of central and Western China defies agricultural skill, and it is in these regions that a surprisingly varied flora obtains. These regions are very sparsely populated, are difficult of access, and, until comparatively recently, were totally unknown to the outside world.

The botanical collections of the two French Roman Catholic priests, les Abbés David and Delavay, of the Russian traveller, N. M. Przewalski, and of the Imperial Maritime Customs officer, Augustine Henry, gave the first true insight into the extraordinary richness of the flora of central and Western China. Delavay's collection alone amounted to about 3000 species, and Henry's exceeded this number! Botanists were simply astounded at the wealth of new species and new genera disclosed by these collections. An entirely new light was thrown on many problems, and the headquarters of several genera, such as, for example, Rhododendron, Lilium, Primula, Pyrus, Rubus, Rosa, Vitis, Lonicera, and Acer, heretofore attributed elsewhere, was shown to be China.

This extraordinary wealth of species exists, notwithstanding the fact that every available bit of land is under cultivation. Below 2000 feet altitude the flora is everywhere relegated to the roadsides, the cliffs, and other more or less inaccessible places. It is impossible to conceive the original floral wealth of this country, for obviously many types must have perished as agriculture claimed the land, not to mention the destruction of forests for economic purposes.

In order to summarize the account of this wonderful flora

THE WEEPING WILLOW (SALIX BABYLONICA) 80 FT. TALL, GIRTH 12 FT.

it is convenient to divide the region into altitudinal zones or belts. The mountainous nature of the country lends itself admirably to such an arrangement, and it is perhaps the only feasible way of dealing with a subject so vast and unwieldy. The chart (p. 7) represents an ideal section of the region and may possibly convey a clearer idea of the subject than the text which follows :—

Division 1.—" The belt of cultivation—2000 feet altitude." The climate of the Yangtsze Valley, up to 2000 feet altitude, is essentially warm-temperate. Rice, cotton, sugar, maize, tobacco, sweet potatoes, and legumes are the principal summer crops ; in winter, pulse, wheat, rape, hemp, Irish potato, and cabbage are generally grown. It is a region of intense cultivation and the flora is neither rich nor varied. The following wild plants are characteristic : Bamboos (*Bambusa arundinacea, Phyllostachys pubescens*, and other species), Fan Palm (*Trachycarpus excelsus*), " Pride of India " (*Melia Azedarach*), Crêpe Myrtle (*Lagerstrœmia indica*), Winter Green (*Xylosma racemosum*, var. *pubescens*), Chinese Banyan (*Ficus infectoria*), Gardenia (*Gardenia florida*), Roses (*Rosa lævigata* and *R. microcarpa*), Nanmu (*Machilus nanmu* and other species), Pine (*Pinus Massoniana*), Soap tree (*Gleditsia sinensis*), Alder (*Alnus cremastogyne*), Privet (*Ligustrum lucidum*), *Paulownia Duclouxii*, oranges, peaches, and other fruit trees, ferns, especially *Gleichenia linearis*, weeds of cultivation, miscellaneous shrubs and trees, including *Pterocarya stenoptera*, Celtis spp., *Cæsalpinia sepiaria*, Wood Oil (*Aleurites Fordii*), and Cypress (*Cupressus funebris*) ; the last two occurring particularly in rocky places.

Division 2.—" Rain forests belt—2000 to 5000 feet altitude." Between 2000 and 5000 feet are found rain forests, consisting largely of broad-leaved evergreen trees, mainly Oak, Castanopsis, Holly, and various *Laurineæ*. The latter family constitutes fully 50 per cent. of the vegetation in this zone. Ferns, evergreen shrubs, Chinese Fir (*Cunninghamia lanceolata*), and Cypress are other prominent components. This belt is interesting also as being the home of nine-tenths of the monotypic genera of trees that are so prominent a feature of the Chinese flora. The more interesting of these are : Eucommia, Itoa, Idesia, Tapiscia, Sinowilsonia, Platycarya, Davidia, Carrieria,

Pteroceltis, and Emmenopterys. Cultivation is less general in this region, and the winter crop especially is of less importance. The crops are similar to those of the belt below except that maize is the staple and displaces rice. In Hupeh this zone is much less extensive and can hardly be said to exist when comparison with its development in western Szechuan is made.

Division 3.—" Cool-temperature belt—5000 to 10,000 feet altitude." From 5000 to 10,000 feet is the largest and most important zone of all. It is composed principally of deciduous flowering trees and shrubs characteristic of a cool-temperate flora and belonging to familiar genera. To these must be added forests of Conifers and many ornamental tall-growing herbs. It is in this zone that is found the astonishing variety of flowering trees and shrubs so pre-eminent a feature of this flora : of Clematis 60 species are recorded from China ; of Lonicera, 60 ; of Rubus, 100 ; of Vitis, 35 ; of Evonymus, 30 ; of Berberis, 50 ; of Deutzia, 40 ; of Hydrangea, 25; of Acer, 40; of Viburnum, 70 ; of Ilex, 30 ; of Prunus, 80 ; of Senecio, 110 ; and the enumeration might be further extended. Pyrus (including Malus, Sorbus, Micromeles, and Eriolobus) is a prominent family in the belt; and behaves in China in the same manner as Cratægus does in the United States of America.

Amongst such botanical wealth it is difficult to make selections, but if any one genus has outstanding claims it is Rhododendron. As in the Himalayan region, so in Western China, the Rhododendrons are a special feature. The genus is the largest recorded from China, no fewer than 160 species being known. I, myself, have collected about 80 species and have introduced upwards of 60 into cultivation. Rhododendrons commence at sea-level, but do not become really abundant until 8000 feet is reached. They extend up to the limits of ligneous vegetation (15,000 feet, *circa*). These plants are gregarious in habit and nearly every species has a well-defined altitudinal limit. In size they vary from alpine plants only a few inches high to trees 40 feet and more tall. The colour of their flowers ranges from pure white, through clear yellow to the deepest and richest shades of scarlet and crimson. In late June they are one mass of colour, and no finer sight can possibly

CHART ILLUSTRATING ZONES OF VEGETATION

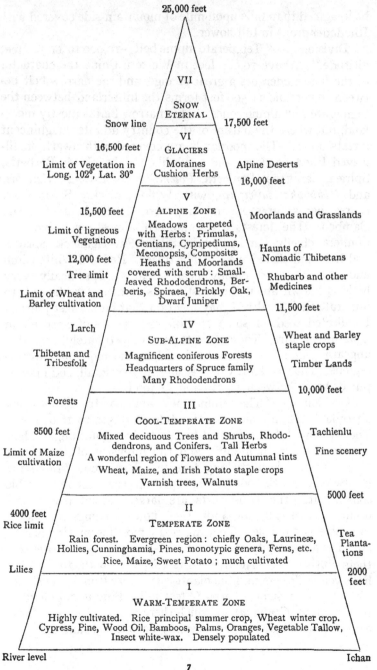

be imagined than mile upon mile of mountain-side covered with Rhododendrons in full flower.

Division 4.—" Temperate alpine belt—10,000 to 11,500 feet altitude." Above 10,000 feet in Western China the character of the flora undergoes a great change, and the narrow belt between 10,000 and 11,500 feet forms the hinterland between the temperate and alpine zones. This narrow belt is mostly moorland, but where the nature of the country admits, magnificent forests occur. The moorlands are covered with dwarf, small-leaved Rhododendrons and scrub-like shrubs, chiefly Berberis, Spiræa, Caragana, Lonicera, *Potentilla fruticosa*, *P. Veitchii*, and *Hippophæ salicifolia*, with Willow, prickly Scrub Oak, coarse herbs, grasses, and impenetrable thickets of dwarf Bamboo. The forests are composed almost exclusively of Conifers, chiefly Larch, Spruce, Silver Fir, Hemlock Spruce, and here and there Pine. A few trees of Red and White Birch and Poplar occur, chiefly near streams. Specifically very little is known about the constituents of these forests, but, to illustrate their wealth, I may mention that on my last journey I collected seeds of some 16 different species of Spruce and 5 of Silver Fir. These forests are, unfortunately, fast disappearing, and are only to be found in the more inaccessible regions. The tree-limit varies according to rainfall, and may be put down as between 11,500 and 12,500 feet.

Division 5.—" The alpine belt—11,500 to 16,000 feet altitude." The alpine zone extends from 11,500 to 16,000 feet. The wealth of herbs in this belt is truly astonishing. Their variety is wellnigh infinite, and the intensity of the colour of the flowers is a striking feature. The genus *Pedicularis* (Louseworts), with 100 species, is perhaps the most remarkable constituent. The Louseworts are largely social plants and occur in countless thousands, their flowers being all colours save blue and purple. They are really most fascinating plants, and it is a great pity that their semi-parasitic nature prevents their cultivation. The Ragworts (*Senecio*), with 100 species, have yellow flowers, and the plants vary in size from low cushion-like plants to strong herbs 6 feet tall. Blue is supplied by the Gentians (*Gentiana*), of which there are 90 species. These again are social plants, and on sunny days the ground for miles

A FOREST OF SPRUCE (PICEA)

is often nothing but a carpet of Gentian flowers of the most intense blue. The Fumeworts (*Corydalis*), with 70 species, supplies both yellow and blue flowers and cannot be denied a place. Then there are the wonderful alpine Primroses (*Primula*). This family is represented in China by some 90 species, four-fifths of which occur in the west. These, like Gentians, take unto themselves in season large tracts of country and carpet them with flowers. Sometimes it is a marsh, at other times bare rock or the sides of streams. One of the most beautiful is *Primula sikkimensis*. Along the sides of streamlets and ponds this species is as common as the Cowslip in some English meadows. Associated with it is its purple congener *P. vittata*. Other striking species are *P. Cockburniana*, with orange-scarlet flowers, a colour unique in the genus ; *P. pulverulenta*, a glorified *P. japonica*, with flower-scapes 3 to 4 feet tall, covered with a white meal and flowers of a rich purple colour ; and *P. Veitchii*, which is best described as a hardy *P. obconica*. Other striking herbs are *Incarvillea compacta* and *I. grandiflora*, both with large scarlet flowers, and *Cypripedium tibeticum*, a terrestrial orchid with enormous pouches, dark red in colour. Also we have Meconopsis in half a dozen species, including *M. Henrici*, with violet-coloured flowers ; *M. punicea*, with dark scarlet flowers ; and *M. integrifolia*, with yellow flowers 8 inches or more across —possibly the most gorgeous alpine plant extant.

Division 6.—"High alpine belt." The limit of vegetation is about 16,500 feet ; a few cushion-like plants belonging to *Caryophyllaceæ*, *Rosaceæ*, *Cruciferæ*, and *Compositæ*, with a tiny species of Primula and *Meconopsis racemosa* being the last to give out. Above this altitude are vast moraines and glaciers, culminating in perpetual snow. The snow-line cannot be less than 17,500 feet. Although at first sight remarkable, the high altitude of the snow-line is easily accounted for by the dryness of the Thibetan plateau and the highlands to the immediate west.

Having briefly outlined the different altitudinal zones and instanced some of the more striking plants characteristic of each, it may be of interest to point out the more important absentees. In China there is no Gorse (*Ulex*), Broom (*Cytisus*),

Heather (*Erica*), nor Ling (*Calluna*) ; the Rock-rose family (*Cistus* and *Helianthemum*) is also unrepresented. The place of Gorse and Broom is inadequately taken by Forsythia, Caragana, Berberis, and various Jasmines ; that of Heather by dwarf, tiny-leaved Rhododendrons, of which there are a dozen or more species. The Cistus family has no representative group unless Hypericum be considered its substitute.

There is practically no pasture-land in central and Western China, but such open country as would compare with commons in England is covered with bushes of Berberis, Spiræa, *Sophora viciifolia*, Caragana, Pyracantha, Cotoneaster, Philadelphus, Holly, and various Roses. The anomalous conditions obtaining in the river-valleys of the west and the peculiar flora found there have been described in Vol. I, Chapter XII.

Another interesting fact, and one that has peculiar reference to the flora of western Hupeh, is the number of plants bearing the specific name *japonica*, which are only Japanese by cultivation and are really Chinese in origin. The following well-known plants are examples : *Iris japonica, Anemone japonica, Lonicera japonica, Kerria japonica, Aucuba japonica, Senecio japonicus,* and *Eriobotrya japonica*. Possibly some of these (and there are many more) may be common to both countries, but I am convinced that when the subject is properly worked out, it will be found that fewer plants are common to both countries than is generally supposed to be the case.

The Chinese flora is largely peculiar to the country itself, the number of endemic genera and species being remarkable even when the size of the country is given due consideration. Yet, in spite of its generally local character, the Chinese flora presents many interesting problems in plant distribution. Not the least interesting is to account for the presence of a species of Libocedrus (*L. macrolepis*), seeing that the other members of this genus are found in California, Chili, and New Zealand. Another noteworthy feature is a species of Osteomeles (*O. Schwerinæ*), which occurs in the far west of China, the other member of this family being found scattered through the islands of the Pacific Ocean. But perhaps the most extraordinary fact in this connexion is the presence on Mount Omei of a species of Nertera (*N. sinensis*), the other members of this

THE THIBETAN MOCCASIN FLOWER (CYPRIFEDIUM TIBETICUM)

family being purely insular and confined to the Southern Hemisphere.

The affinity of the Chinese flora, with contiguous and distant countries, is an interesting theme and one that could be enlarged upon at length. The Himalayan flora is represented by certain species in Western and central China, and there is a considerable affinity between the floras of these regions. This is to be expected, yet it presents problems of exceptional interest, since it is the Sikhim element which comes out strongest. When the flora of Bhutan and of the country between Bhutan and Western China is properly explored it will probably be found that Sikhim represents the most western point of distribution for certain plants rather than their real headquarters. Of Himalayan plants commonly met with in the region, with which this work is intimately concerned, the following examples may be given : *Evonymus grandiflora, Euptelea pleiosperma, Clematis montana, C. grata, C. gouriana, Rosa sericea, R. microphylla, Primula sikkimensis, P. involucrata, Podophyllum Emodi,* and *Amphicome arguta.* In Yunnan there is a decided affinity with the Malay-Indian flora.

The aggressive nature of the Scandinavian (British) flora is evidenced by the following herbs and shrubs which are locally very common : Vervain (*Verbena officinalis*), Agrimony (*Agrimonia Eupatoria*), Buttercups (*Ranunculus acris, R. repens,* and *R. sceleratus*), Silver-weed (*Potentilla anserina*), Great Burnet (*Poterium officinale*), False Tamarisk (*Myricaria germanica*), Ivy (*Hedera Helix*), Bird Cherry (*Prunus Padus*), and Plantain (*Plantago major*).

In the north and throughout the upland valleys and high-lands of the west a few Central Asian and Siberian forms occur, such as *Sibiræa lævigata, Spiræa alpina, Cotoneaster multiflora, Thalictrum petaloideum, Delphinium grandiflorum,* and *Lonicera hispida.*

At first sight it would very naturally be supposed that the Chinese flora was most closely allied if not to that of Europe at least to that of the Asiatic continent generally. Yet this is not so. The real affinity is with that of the Atlantic side of the United States of America ! This remarkable fact was first demonstrated by the late

Dr. Asa Gray when investigating the early collections made in Japan. Modern work in China, and especially central China, has given overwhelming evidence and established beyond question Asa Gray's conclusions. There are many instances in which only two species of a genus are known—one in the eastern United States and the other in China. Noteworthy examples are the Tulip tree, Kentucky Coffee tree, the Sassafras, and the Lotus Lily (*Nelumbium*). A considerable number of families are common to both countries, and in most instances China is the dominant partner. Usually the U.S.A. have one and China several species of the same genus, but here and there the opposite obtains. Magnolias afford a good illustration of this affinity. This genus, absent from Europe and western North America, is represented by 7 species on the Atlantic side of the North American continent, and by 19 species in China and Japan.

The following brief list still further illustrates this :—

SOME GENERA COMMON TO CHINA, JAPAN, AND THE ATLANTIC SIDE OF THE UNITED STATES OF AMERICA

CHINA AND JAPAN		UNITED STATES OF AMERICA	
Genus	*No. of Species*	*Genus*	*No. of Species*
Magnolia	19	Magnolia	7
Schisandra	10	Schisandra	1
Itea	5	Itea	1
Gordonia	3	Gordonia	2
Hamamelis	2	Hamamelis	2
Shortia	3	Shortia	1
Catalpa	5	Catalpa	2
Negundo (Acer)	5	Negundo (Acer)	1
Wisteria	4	Wisteria	2
Astilbe	10	Astilbe	1
Podophyllum	6	Podophyllum	1
Illicium	6	Illicium	2
Stewartia	2	Stewartia	2
Berchemia	8	Berchemia	1
Nyssa	1	Nyssa	4

In a few cases the same species is common to both countries. The most extraordinary instance of this is *Diphylleia cymosa* (Umbrella Leaf). This plant occurs in localities separated by 140° of longitude and exhibits absolutely no marked variation.

In the instances mentioned above, the families are absent

from any other region in the world. In others,—for example, Oak, Hornbeam, Elm, Birch, Ash, Beech, and Sweet Chestnut, —where the families range around the whole temperate zone of both Old and New Worlds, the individual Chinese species are usually more closely akin to those of North America than to those of Europe.

The explanation of this phenomenon is to be found in the glaciation of the Northern Hemisphere in prehistoric times. In those far-off times the land connexion between Asia and North America was far more complete than it is to-day, and the flora extended much farther to the north. The ice-cap which gradually crept down forced the flora to travel towards the equator. Later, when the period of great cold was over, and the ice-cap receded, the plants crept back ; but the ice-cap remained at a more southern latitude than before, and consequently rendered much of the land formerly covered with forests too cold to support vegetable life of any sort. This rearrangement after the Ice Age caused a break between the two hemispheres, and the consequent isolation and cutting off of the floras. Other agencies and factors played a part, but the above explains briefly and roughly why the floras so much alike should to-day be so widely separated geographically.

That the Chinese flora is an ancient one is evidenced by the number of old types it contains. For example, in ancient times, *Ginkgo biloba* (Maidenhair tree) was found, not only in Asia, but in Western Europe, Northern California, and Greenland, as the fossil remains found in Jurassic beds of these countries testify. To-day it exists only in China and Japan as a cultivated tree, being preserved to us by the Buddhist and other religious communities who plant it in the neighbourhood of their temples. Cycas, Cephalotaxus, Torreya, and Taxus are other old types, but these occur in a wild as well as in a cultivated state in China to-day. Many of the older ferns, such as Osmunda, Gleichenia, Marattia, and Angiopteris, are common in China and widely spread. In speaking of the older ferns it may be of interest to note that Augustine Henry discovered in Yunnan an entirely new genus of *Marattiaceæ*, which has been named Archangiopteris.

From the evidence before us it would appear that the Chinese flora suffered less during glacial times than did that of Europe and North America. This may possibly have been due to the greater continuity of land towards the equator which obtains in Asia as compared with that of the continents of Europe and America.

CHAPTER II

THE PRINCIPAL TIMBER TREES

THE forested regions of China are to-day remote from the populous parts of the country, and are only to be found in the more wildly mountainous parts, which are little suited to agriculture, and where the rivers are unnavigable rock-strewn torrents, and roads, as such, can scarcely be said to exist. Such districts are always at considerable elevation and are but sparsely peopled. In all the more accessible regions agriculture has claimed the land, and the trees are only met with around houses, temples, tombs, stream-sides, or crowning cliffs. The scarcity of timber is acutely felt throughout the length and breadth of the land. Dressed logs and poles are carried long distances to navigable waterways and floated either down or up-stream, consequently their cost is high. The ports on the sea-board and lower Yangtsze import timber in quantity for general construction purposes from Puget Sound and British Columbia. A certain amount also comes from Japan. Hardwoods for miscellaneous purposes are imported from various parts of Malaysia, and a certain amount of Jarrah wood for railway work has recently been sent from Australia. The famous blackwood furniture of China is not made of native wood, but of timber imported from Bangkok, Saigon, and other places in Indo-China. Botanically the source of " Chinese blackwood " is unknown. The so-called " Bombay blackwood " is derived from *Dalbergia latifolia*, and possibly the " Chinese " kind is from a closely allied species. Western China is rather better off for timber than other parts of China, and fortunately so, since the importation of timber as a business is utterly impossible. Nevertheless there is a great dearth of wood for building purposes, and timber prices have doubled during the last decade. The massive timbers to be seen in old Chinese temples and houses are now unobtainable from the native trees of China.

Since the scarcity of timber is so great, every kind of tree found in the thickly populated regions furnishes wood of some value, but for the purpose of this chapter it suffices to give a brief account of the more important kinds and those most generally useful.

By far the most important "timber" in China is, of course, the stems of the Bamboo. The Jesuit priest, Trigault, in a work on China, published in 1615, states : " They have a kind of reed called *Bambu* by the Portuguese. It is almost as hard as iron. The largest kind is scarcely encompassed with two hands. It is hollow inside and presents many joints outside. The Chinese use it for pillars, shafts of lances, and for 600 other domestic purposes."

Although three centuries have elapsed since the above quotation was written it applies equally to the conditions of the present day, for the uses to which the Bamboo is put in China are indeed limitless. It supplies many of the multifarious needs of the people with whose everyday life, from birth to death, it is inseparably entwined. From Bamboo stems are fashioned the various household utensils, furniture, the house itself, many agricultural implements, masts and gear for boats, rafts, ropes, bridges, irrigation-wheels, water-pipes, gas-pipes, tubes for raising brine, sedan-chairs, tobacco and opium-pipes, bird-cages, snares for entrapping insects, birds, and animals, umbrellas, raincoats, hats, soles for shoes, under-shirts, sandals, combs, musical instruments, ornamental vases, boxes, and works of art, the pen (brush) to write with, the paper to write upon, everything, in fact, useful and ornamental, from the hats of the highest officials to the pole with which the coolie carries his load. Formerly the records of the race were written on bamboo tablets which were strung together at one end like a fan. Records of this description, dug up in A.D. 281, after having been buried for 600 years, were found to contain the history of Tsin from 784 B.C., and incidentally also that of China for 1500 years before that date.

THE SPINY BAMBOO (BAMBUSA ARUNDINACEA) 60 FT. TALL

Bamboo shavings are used in caulking boats and for stuffing pillows and mattresses. The young shoots are a valued vegetable. According to popular belief, in times of scarcity a compassionate Deity causes the Bamboo to flower and yield a harvest of grain to save the people from starvation. The Bamboo flourishes everywhere in the Far East, and is just as beautiful when sheltering the peasant's cottage or beggar's hut as when ornamenting the courtyards of temples and the mansions of the wealthy. It is the one woody plant that is really abundant throughout all but the coldest parts of the Middle Kingdom. The Occident possesses no tree or shrub which for all-round general usefulness compares with the Bamboo of the Orient.

The Chinese generic name for the Bamboo family is "Chu," the different kinds being distinguished by a prefix. The natives have no difficulty in recognizing the various species, but botanists generally have found Bamboos exceedingly difficult to deal with systematically. In the *Index Floræ Sinensis* 33 species are enumerated, but for the purpose of this chapter only 4 or 5 species are involved.

Throughout the Yangtsze Valley, up to about 2500 feet altitude, the " Pan chu " (*Phyllostachys pubescens*) is one of the commonest species. Its young spear-like stems rear themselves 30 to 40 feet, and finally develop into beautiful arched plumes. The stems are about 3 or 4 inches in diameter, dark shining green, becoming yellow with age. The wood is moderately thick and is used for a great variety of purposes. It is largely employed on the Yangstze, above Ichang, for making tracking lines for the various river craft. A species allied to this, but smaller in every way, never exceeding 20 feet in height, is the Ch'ung chu (*P. heteroclada*). This Bamboo is commonly used in western Hupeh for paper-making.

A very common species in the warmer parts of Szechuan is the "Tz'u chu" (*Bambusa arundinacea*, often called *B. spinosa*), the Spiny Bamboo. This magnificent species produces stems 50 to 75 feet tall and 8 to 10 inches in diameter at base. It does not spread very much, but forms compact clumps, which are impenetrable on account of their density and the presence of innumerable, slender, ferociously spiny stems which develop among and around the larger culms. This Bamboo has a small core and very thick wood. It is used in household carpentry, for furniture, ornamental vases, boxes, and scaffolding, and has a hundred and one other uses. Another species is the Nan chu (*Dendrocalamus giganteus*), the largest growing of all the Bamboos found in western Szechuan. This is confined to the warmer parts of the province, where it forms wide-spreading groves. The stems grow 60 to 80 feet tall and are 10 or 12 inches thick. The core is very large, the wood thin and light. It is commonly used for constructing the rafts which ply on the shallow but turbulent rivers of western Szechuan. It has also many other uses and is especially prized for making chop-sticks.

Yet another very commonly cultivated species is *Bambusa vulgaris*, sometimes called the Kwanyin chu, which produces pale-coloured stems 30 to 50 feet tall. The wood is thin and is used for a variety of purposes, but is less valuable than any of the foregoing. The young shoots of these large-growing Bamboos are cut just as they appear above the ground, and eaten as a vegetable, the flesh being white, firm and crisp.

Apart from Bamboo the most common timber for all-round use is that derived from the " Sha shu," or " China Fir" (*Cunninghamia lanceolata*). This coniferous tree is widely spread throughout warm-temperate parts of China and is especially partial to red sandstone. It is particularly abundant in the Yachou Prefecture and on the mountains bordering the north-west corner of the Chengtu Plain. It grows from 80 to 120 feet tall, and has a straight mast-like stem ; after the trees are cut down this Conifer reproduces itself by sprouts from the old stumps. The bark is commonly employed for roofing purposes. The wood is light, fragrant, and easily worked. For general building purposes, house-fittings, and indoor carpentry it is the most esteemed of all Chinese timbers ; also it is in great request for coffin-making, the fragrant properties of the wood being considered to act as a preservative. For ordinary coffins several logs are dressed and fastened together laterally to form a thick, wide plank called " Ho-pan," four of which, with two end pieces added, make a coffin. All who can afford it have such coffins

THE CHINESE FIR (CUNNINGHAMIA LANCEOLATA) 120 FT. TALL,
GIRTH 20 FT.

lacquered a shining jet-black. But the more expensive
coffins are those in which each Ho-pan is hewn from a single
log of timber, and the most valuable of all are those made
from Hsiang Mu (fragrant wood), or Yin-chên Mu (long-
buried wood). For such a coffin 400 to 1000 ounces of silver
is the usual price. For the most part, Yin-chên Mu comes
from the Chiench'ang Valley, where it was probably engulfed
as the result of an earthquake in times past. In 1904 I
ascended the Tung Valley from Fulin to Moshi-mien, *en route*
for Tachienlu, and near the hamlet of Wan-tung came upon
a place where natives were engaged in excavating buried
timbers. The work was being carried on in a narrow valley.
At the head of the valley a torrent had been dammed up
and the accumulated waters, released at will from time to
time through a sluice, carried much of the overlying debris
away. Many of the excavations were fully 50 feet deep.
All sorts of timber is found buried in this place, but only
the " Hsiang Mu " (fragrant wood) is considered of value. I
procured a specimen of this wood, and subsequent microscopic
examination has proved it to be that of *C. lanceolata*. The
Chinese consider that these trees have been buried for
two or three hundred years. The timber is wonderfully
preserved and is more compact in texture and more fragrant
than that of recently felled trees. Ho-pans made from
" Hsiang Mu " average about 30 inches wide and 7 feet in
length. In all my travels in Western China I have seen
only one living specimen of Cunninghamia approaching to
the size of these long-buried giants.

In Chengtu and neighbouring cities, the timber known as
" Lien sha," derived from *Abies Delavayi* and allied species, is
generally employed for all the larger beams, pillars, and planking
in house-building. This handsome Silver Fir (*Abies Delavayi*)
is common on all the higher mountains of the west, but that
growing in the Yachou prefecture is most accessible, and this
district is the main source of the timber-supply. The timber is
soft and not very durable, but the large size of the logs render
it most serviceable. The Pine (*Sung shu*) is very common, the
most widely distributed species being *Pinus Massoniana*. This
tree ascends from sea-level to 4000 feet altitude. The timber

obtained from the higher altitudes is close-grained, resinous, and durable, but that from low-levels is soft, very open, and of little value. Other Hard Pines, such as *P. Henryi, P. densata, P. Wilsonii, P. prominens*, are found at higher altitudes (up to 10,000 feet) and yield valuable timber, but unfortunately they occur only in inaccessible places. The Chinese White Pine (*P. Armandi*) is widely spread in the more mountainous parts. This tree never attains any great size, but the timber is very durable and resinous. It is esteemed for building purposes and for making torches.

All the Conifers yield useful timber, but unfortunately few are found to-day in accessible regions. Around Tachienlu the Hung sha (Red Fir), *Larix Potaninii*, is esteemed the most valuable of all timbers. The Tieh sha (*Tsuga yunnanensis, T. chinensis*) is made into shingles for roofing purposes and is also valued for planking. In the Lungan prefecture the Mê-tiao sha (*Picea ascendens*) is a most valuable timber for general building purposes. Many other kinds of Spruce (*Picea*) occur on the mountains, and with Silver Fir (*Abies*) and Larch (*Larix*) form the only remaining Conifer forests in Western China. Juniper (Hsiang-peh sha), *Juniperus saltuaria*, is common north of Sungpan, where it is valued for building purposes. *Cupressus torulosa* (K'an-peh sha) occurs in the arid valleys of the west ; *Taxus cuspidata*, var. *chinensis* (Tuen-ch'u sha), and *Keteleeria Davidiana* (Yu sha or Oil Fir) are found scattered all over Western China between 2000 and 5000 feet altitude, but are nowhere really abundant.

From Ichang westward, up to 3500 feet altitude, the commonest Conifer next to Pine is the Peh sha or White Fir (*Cupressus funebris*), and in the more rocky limestone regions it is the more common tree of the two. This handsome Cypress, with its pendant branches, is generally planted over tombs and shrines and in temple grounds. The wood is white, hard, heavy, and exceedingly tough. It enters largely into the structure of all boats plying on the Upper Yangtsze, forming the sides, bulkheads, and often the cross-beams and decks. It is also made into chairs, tables, and other furniture. The superstructure of the boat is usually of Sha Mu (*Cunninghamia*), the bottom and main timbers of Oak and Nanmu.

THE CHINESE PINE (PINUS MASSONIANA) 80 FT. TALL, GIRTH 10 FT.

Oak is widely dispersed from river-level to 8000 feet altitude, but large trees are scarce except in the vicinity of tombs, shrines, and other sacred places. A general name for the family is "Li," and the Chinese distinguish many kinds, such as Peh-fan, Hwa, Hung, Tueh, and Chu li ; botanically about a score of species occur in this region, of which the commonest are *Quercus serrata*, *Q. variabilis*, and *Q. aliena*. All yield close-grained timber, highly valued for a variety of purposes apart from boat-building.

Nanmu (Southernwood) includes a number of species of Machilus and Lindera. All are evergreen and singularly handsome trees. They are largely planted around homesteads and temples in Szechuan, and are a prominent feature of the scenery of parts of the Chengtu Plain and around the base of Mount Omei. They grow to a great size and have clean, straight trunks and wide-spreading, umbrageous heads. The timber is close-grained, fragrant, greenish and brown in colour, easily worked, and very durable. It is highly esteemed for furniture-making, and for pillars in the temples and the houses of the wealthy. As planking it is used for boat bottoms. Nanmu is one of the most valuable of all Chinese timbers, and the tree itself among the handsomest of evergreens. Camphor (Ch'ang shu), *Cinnamomum Camphora*, is found scattered over western Hupeh and Szechuan up to 3500 feet altitude, and its fragrant timber, like that of Nanmu, is made into high-class furniture. The wood furnished by the thick main roots of this tree is known as "Ying Mu" and is valued for cabinet work.

For high-grade cabinet work, picture frames, and the very best furniture the timber most highly esteemed in Szechuan is the "Hung-tou Mu," derived from *Ormosia Hosiei*, a tree allied to the Sophora. In the spring *O. Hosiei* produces large panicles of white and pink pea-shaped flowers, and at all seasons of the year is a striking tree. The wood is heavier than water, of a rich red colour, and beautifully marked. It is the most high-priced of all local timbers, and is now very scarce. In north-central Szechuan it is still fairly common, but on the Chengtu Plain it is only found in temple grounds or over shrines. The native name signifies "Red Bean tree," the seeds being red and contained in bean-like pods. Allied to the foregoing is

Dalbergia hupeana, which yields the valuable " T'an Mu," a wood whitish in colour, very heavy, and exceedingly hard and tough. It is almost exclusively employed in building the wheel-barrows used on the Chengtu Plain; for the handles of carpenters' tools, rammers for oil-presses, blocks and pulleys used on boats, and for every purpose where stress and strain obtain. This tree grows tall (80 feet) but is never of any great thickness ; it is widely spread in the west up to 3000 feet altitude.

Three other members of the Pea family that yield useful woods of greater or less value, are the " Huai shu " (*Sophora japonica*, " Tsao-k'o shu " (*Gleditsia sinensis*), and " Yeh-ho shu " (*Albizzia lebbek*). All three species are common, the first two forming a characteristic feature of the vegetation of the more arid river-valleys of the west. The wood of these trees is used in general carpentry and furniture-making.

One of the commonest trees throughout the hot, rather arid river-valleys, up to 8500 feet altitude (but by no means confined thereto), is *Juglans regia*, the Walnut (Hei-tou shu). It is cultivated for its fruits, which are a valued article of food and a source of oil. The wood has recently become in great demand in the newly established arsenals for making rifle-stocks. The supply is not equal to the demand, and much Nanmu timber is used as a substitute. This latter is lighter and less serviceable for this purpose than that of the Walnut.

The best rudder-posts are made from the wood of the " Huang-lien shu " (*Pistacia chinensis*), a large-growing tree found everywhere up to 5000 feet altitude. A log having a natural " fork " at one end is in general use for the balance-rudder on all the larger boats. The wood of the Loquat (Pi-pa shu), *Eriobotrya japonica*, which is red-coloured, heavy, and of great strength, is also employed for this purpose. The young shoots of the Pistacia, known as " Huang-ni ya-tzu," are cooked and eaten as a vegetable, and so also are the shoots of the " Ch'un-tuen shu " (*Cedrela sinensis*). This last-named tree furnishes a valuable timber, beautifully marked with rich red bands on a yellowish-brown ground. Foreigners call it Chinese Mahogany. It is easily worked, does not warp nor crack, and is esteemed for making window-sashes, door-joists, and furniture. The tree grows 80 feet tall, the trunk is very

A DRESSED LOG OF HEMLOCK SPRUCE 18 FT. 6 IN. BY 9 IN. BY 7 IN.

straight, and but little branched. It is quite common in western Hupeh up to 4500 feet altitude, but much less so in Szechuan.

Tea-chests for all the higher-grade teas are made of wood derived from the Chinese Sweet Gum (Feng-hsiang shu), *Liquidambar formosana*. This is a strikingly handsome tree, growing 80 to 100 feet tall, with a girth of 12 to 15 feet. It occurs scattered all over the west up to 3500 feet altitude ; the leaves turn a rich red-brown in autumn and remain on the trees far into the winter.

The best carrying poles are made from the " Tzu-k'an shu " (*Ehretia acuminata* and *E. macrophylla*), the wood of these trees being light but very tough. Oak and Bamboo are also used for the same purpose and are cheaper. For making the drums used on boats and in temples the wood of the " Tzu-ch'in shu " (*Kalopanax ricinifolium*) is considered best, being easily worked, pliable, and resonant. The two ends of the drum are covered with hide.

The finest Joss-sticks (Chinese incense) are composed of the pounded leaves and branches of various members of the Laurel family (*Laurineæ*), all of which are rich in fragrant, essential oils. As an adulterant the pulped wood of Cypress and Birch is commonly employed.

On the barren hills around Ichang and elsewhere the Common Pine (Sung shu), *Pinus Massoniana*, has been planted as a source of fuel. Along the stream-sides and canals on the Chengtu Plain, Alder (Ching shu), *Alnus cremastogyne*, is generally planted for the same purpose. The Alder and Pine, together with Bamboo, are the only trees planted for the economic value of their timber. On the mountains, Beech, Ash, Poplar, Sweet Chestnut, Hornbeam, Birch, and many other valuable and useful timber trees occur, but are difficult of access and consequently not in general use.

CHAPTER III

FRUITS, WILD AND CULTIVATED

CHINA is the original home of several fruits which are now cultivated all over the world, as, for example, the orange, lemon, pomelo, peach, and Japanese plum. In the south a number of tropical fruits, such as banana, pineapple, papaw, areca-nut, litchi, longan, and "Olives" (*Canarium*), are grown, but only the last three, and these in very small quantities, are found in the regions with which we are concerned. In the north, more especially around Chefoo, apples and pears, introduced from America, are cultivated and very excellent fruit is produced. In the north, too, very fine grapes are grown, and the fruit generally is of a high order. But, in general, little attention is given to fruit-culture; pruning the trees and thinning the fruit is not attended to, with the result that nearly all Chinese fruit is lacking in quality. Usually it is gathered before it is properly ripe, and this has much to do with the absence of flavour which is unfortunately characteristic. Particularly is this indifference and neglect evident in central and Western China, where a very considerable quantity and variety is grown. The oranges, peaches, and persimmons are equal to those obtainable anywhere, but all the other succulent fruits are of very low-grade quality. It is to be regretted that more attention is not given to the subject, for the region could undoubtedly be made to produce the very best of fruits.

In ascending the Yangtsze River, from where the foothills commence below Ichang, and westward to Sui Fu, Orange-groves are a feature, attaining their greatest luxuriance between Chungking and Lu Chou. In December, when the trees are laden with ripe fruit, these groves are a remarkable sight. The Orange is happiest when growing on the leeside of rocky

THE MANDARIN ORANGE IN FULL FRUIT

escarpments, or at their base, where it is protected from the winds. It is very partial to the clayey marls and sandstones of the Red Basin. In western Szechuan the loose-skinned or Mandarin Orange (*Citrus nobilis*) is most generally grown. In season the fruit can be purchased on the spot at the rate of 500 to 1000 for a shilling. Unfortunately this orange does not keep well, but when removed and dried the rind constitutes a favourite medicine known as " Chien-yün-p'i." The fibres and pithy substance surrounding the fleshy carpels within the rind also form a medicine which is called " Chü-lo." In the gorges a tight-skinned or Sweet Orange, " Shan K'an-tzu " (*C. Aurantium*, var.), is more usually met with. The so-called Ichang orange of this type is noted far and wide in China. It has a higher market-value than the " Mandarin " and keeps well. In Chengtu these oranges are kept fresh and good all through the summer, but by what process I failed to discover.

A Lemon (*C. ichangensis*) is also grown in the Ichang Gorge, but is not common. The fruit of this species is broadly oval in shape and of excellent flavour. Pomelos, " Yō-tzu " (*C. decumana*, var.), are met with, but the fruit seldom contains any pulp worthy of the name, consisting usually of little but pith and seeds. The Kumquat (*C. japonica*) is sparingly cultivated for its fruits, which, preserved with sugar, are an esteemed delicacy. A Citron (*C. Medica*, var. *digitata*) is also occasionally grown for its curious-looking fruit which is known as " Fingered Citron," or " Buddha's Hand."

The Orange and allied fruit trees are propagated by notching the shoots which arise from the base of the tree and fixing earth around the cut. A framework of bamboo or a broken earthenware pan is used to keep the soil in place. When many roots have been formed in the heaped-up soil a final severance of the shoot from the parent tree is made, and in due course the new plant is removed to a permanent site. Boring-insects are unfortunately making sad havoc among the Orange-groves in Western China. No attempt at prevention or control is made by the owners, and nothing but the wonderful vitality of the tree saves it from extinction.

The Peach, " Tao-tzu " (*Prunus Persica*), is abundantly

cultivated in Hupeh and Szechuan from river-level to 9000 feet altitude. Freestone and clingstone varieties and oval and flattened kinds occur ; those from the vicinity of Ichang are of delicious flavour and are probably not excelled anywhere in the world. The climate more than anything else is responsible for this, since the trees are little cared for and generally covered with the San José scale-insect. The trees are grown in orchards or in small groups around houses, but sub-spontaneous bushes are met with everywhere by the wayside and on cliffs. An oil is extracted from the kernels in northern China, but not in the western parts of the Empire, as far as my observation goes. The Peach was introduced into Asia Minor and Europe from Persia somewhere about 300 B.C., but it has been cultivated in China from very remote times and was probably carried to Persia by way of the old trade route via Bokhara. Whilst it is now accepted that China is the original home of this invaluable fruit, it is by no means certain as to what particular plant represents the wild type. A species found in northern China and known as *P. Davidiana* is generally regarded as the source of origin of the cultivated peach. From this view I, however, dissent. My opinion is that the species are distinct, and that the type of the garden peach is no longer to be found in a wild state. The nearest to it is the sub-spontaneous form, plants of which are abundant on the cliffs and by the waysides all over western Hupeh and Szechuan. In this connexion it may be of interest to record that in the neighbourhood of Tachienlu I discovered a new species of Peach which has since been named *P. mira*. This plant is a typical freestone Peach in every respect, but has a small, smooth, ovoid stone. It is now in cultivation, and, coming as it does from a very cold climate, may eventually prove the progenitor of a hardier race of cultivated Peaches.

The Apricot (*P. Armeniaca*) is generally supposed to be a native of Armenia, as its name implies, from whence it was introduced to China, where it has long been cultivated ; but Maximowicz regarded it as spontaneous in the mountains near Peking. The Apricot tree grows to a large size (40 to 50 feet), but the fruit, known as " Hun-tzu," is fibrous and very harsh in flavour. There is room for the improved

THE WALNUT (JUGLANS REGIA) 60 FT. TALL, GIRTH 12 FT.

varieties of apricot in China, as the dried apricots prepared in northern India, which find their way across Thibet to Western China, are highly esteemed by Thibetans and Chinese alike.

Plums, " Ku-li-tzu," are commonly cultivated, the fruits being round in shape and either green, yellow, red, or purple in colour, but all are of indifferent flavour. All these cultivated forms are derived from *P. salicina*, a tree common in the thickets and margins of woods throughout Hupeh and Szechuan. Under the name of Japanese Plum this species has been introduced into California, Europe, and elsewhere, and is now widely cultivated. Authentic specimens of the species from which the plums cultivated in Europe have been derived (*P. communis*) have not been recorded from China, and very probably it does not occur there. The Japanese Apricot (*P. mume*), so widely cultivated in China and Japan, where it is dwarfed and trained into curious shapes and much appreciated for its early flowering propensities, is wild in western Hupeh and Szechuan, being known as " Oo-me." The fruit is round, usually red on one side and yellow on the other, of indifferent flavour, and rendered less palatable by its felted, woolly stone.

The Common Almond is not grown in China, but in 1910, near Sungpan, I discovered an allied species, since named *P. dehiscens*, in which the ripe fruit opens and exposes the stone. The kernel of this fruit is eaten and locally is much esteemed. The plant forms a very dense, spiny bush, 5 to 12 feet tall, and is very abundant in the upper reaches of the Min Valley. The fruit may be described as " dry," since hardly any " flesh " is developed. This species is now in cultivation, and is certainly an interesting addition to the Almonds hitherto grown.

Cherries, "Ying-tao," are abundant in the woods and forests and run riot in species. In *Plantæ Wilsonianæ*, Part II, Koehne describes no fewer than 40 species based on material collected by me alone ! The Cherry is, however, rarely cultivated, and such fruit as is on sale at Ichang and elsewhere is small and lacking in flavour. Its chief merit is in being the first stone-fruit of the season, coming into the market the end of April. The Cherry cultivated around Ichang is *P. involucrata.* The species from which the European cherries have been derived (*P. avium* and *P. Cerasus*) are not found in China.

The Pear, "Li-tzu," is very generally cultivated and is especially abundant throughout the upper reaches of the river-valleys in the west. It is also common in the higher parts of the glens which lead off from the gorges in western Hupeh. Several kinds are grown, and in some instances the fruit attains a very large size. Usually these pears are as hard as rock, and though very useful for cooking purposes are of little value for dessert. Propagation by crown-grafting is commonly practised, but little attention is given to the trees afterwards. All the varieties of Chinese pears have been evolved by long cultivation from native species (probably *Pyrus sinensis, P. ussuriensis,* and another species not yet authentically named, but known in China as the " Tang-li "), and have not common origin with those cultivated in the Occident which have been derived from *P. communis.* Around Peking the Chinese cultivate a peculiar kind of Pear under the name of " Peh-li-tzu " (White Pear). The fruit is apple-shaped, about 1¾ inches in diameter, pale yellow in colour, and of most delicious flavour. This pear is probably a superior variety of *Pyrus ussuriensis.*

Apples are much more sparingly cultivated than pears, with which they are grown in association. They are more frequent around Sungpan Ting and Tachienlu than in Hupeh. The fruit is small, green, or greenish-yellow on one side and rosy on the other in the best variety, with an agreeable bitter-sweet flavour. It is uncertain as to what species these apples belong, but possibly to *Malus spectabilis, M. prunifolia*

The Quince, "Mu-kua," is commonly cultivated in central China, but less so in the west. The fruits are oiled and kept as ornaments in houses, being appreciated for the fragrant odour. They are also used as medicine. Two species occur—*Chaenomeles sinensis* with nearly round leaves and dark red flowers and *C. cathayensis* with elongated leaves and white flowers, flushed pink. Closely allied to the quince is *Docynia Delavayi,* which is very abundant in Yunnan, where the fresh fruits known as "Tao yi" are used in ripening persimmons. The fruits of each are arranged in alternate layers in large jars and covered with rice-husks, and in ten hours the persimmons are bletted and fit for eating. The Docynia occurs sparingly in western Szechuan, but in that locality the fruit is not utilized.

The Loquat, "P'i-pa" (*Eriobotrya japonica*), both wild and cultivated, occurs in quantity up to 4000 feet altitude, and is most abundant in rocky places. This handsome evergreen forms a tree 30 feet tall, and produces its fragrant white flowers in the early winter, the fruit being ripe in April. The fruit is orange-coloured, of a pleasant sub-acid flavour, but there is very little "flesh" surrounding the large, soft brown seeds, which have an almond-like taste and might be used for flavouring purposes.

In different parts of China various species of Hawthorn, "Shan-li-hung-tzu" or "Shan-cha," are cultivated for their fruits. In Hupeh the species thus favoured is *Cratægus hupehensis*; orchards of this tree occur in the neighbourhood of Hsingshan Hsien. The fruit is scarlet, nearly 1 inch in diameter, but of insipid flavour.

One of the most delicious of all fruits grown in China is the Persimmon, "Tsze-tzu" (*Diospyros kaki*). The Persimmon tree is abundant up to 4000 feet altitude, and usually forms handsome specimens 60 feet or more tall. The fruit may be ovoid or flattened-round, and with or without seeds. It is not really edible until dead ripe, at which stage all the tannic acid is dissipated or changed into sugar. The Chinese have various methods of ripening this fruit to bring out its full flavour. The process, in the main, consists in stratifying and covering them with rice-husks and admitting only a modicum of air. Persimmons are often allowed to remain on the trees long after the leaves have fallen, and the masses of orange-coloured fruits on such trees present a wonderful sight.

In the neighbourhood Lu Chou are cultivated Litchis (*Nephelium Litchi*) and Longans (*N. longana*) as orchard fruits. They thrive very well in this district, and the fruits command high prices in the market. The Chinese Olive (*Canarium album*) is also grown in the same locality. In the arid river valleys of the west the Chinese Date-plum, "T'sao-tzu"

(*Zizyphus vulgaris*), is frequently cultivated, but the quality of the fruit is very poor, and cannot compare in size and flavour with that produced in Shantung and other parts of north-eastern China. In the warmer parts the Pomegranate, " Tsze-niu " (*Punica Granatum*), is commonly met with, but the fruit is scarcely edible. In Yunnan very fair pomegranates are grown. Although widely spread and naturalized in parts of China, competent authorities consider the Pomegranate to have been introduced there.

Grapes, " Chia-p'u-tao," are sparingly cultivated in the west, but the quality is very inferior to those grown around Peking. The only kind I have seen has white fruit. The varieties commonly cultivated are all forms of *Vitis vinifera*, which, according to Bretschneider, was introduced into China from Western Asia during the second century B.C. Around Kiukiang the Spiny Vitus, " P'u-tao-tzu " (*V. Davidii*), is sometimes cultivated. This vine produces black, globose grapes of good size and appearance, but the flavour is very harsh. It occurs as a common wild plant in the mountains of the west.

The Walnut, " Hei-tao " (*Juglans regia*), is an exceedingly common tree throughout the regions with which we are concerned, ranging up to 8500 feet altitude. It is especially abundant in the arid river valleys of west Szechuan, and equally so in the mountains and valleys of Hupeh. The nuts vary considerably in size, shape, and in the thickness of the shell. The best are of large size and have very thin shells. They are valued not only as a food but for their sweet-oil, which is expressed and used for culinary purposes. A Butternut, " Yeh Hei-tao " (*J. cathayensis*), is also common in the woods and thickets. The kernels are eaten, but the shell is very thick and difficult to crack.

The seed of the Maidenhair tree, " Peh-k'o " (*Ginkgo biloba*), after being roasted is esteemed as a dessert nut. The seed of the Lotus Lily, " Lien hwa" (*Nelumbium speciosum*), Groundnut, " Lao-hua-tsen " (*Arachis hypogæa*), are similarly valued. The Water-chestnut, " Ling-chio " (*Trapa natans*), is abundantly cultivated and the fruit is eaten.

In the woods and thickets many kinds of wild fruits are

THE FLOWERS AND FRUIT OF ACTINIDIA CHINENSIS

found which are eaten locally. Brambles (*Rubus*) in great
variety occur, over 100 species being recorded from China. The
majority yield edible fruit, and in some cases this is superior
to any found elsewhere in the world. I have succeeded in
introducing about 30 species, and look forward to the day when
some one will seriously take up the culture of brambles and by
hybridizing them evolve a new race of berries to add to the
soft fruits at present in cultivation. The three best of the
new introductions, according to my own palate, are *Rubus
pileatus, R. amabilis,* and *R. corchorifolius,* all vinous-flavoured,
raspberry-like fruits. The black fruits of *R. omeiense* and *R.
flosculosus* are also good eating, as are the orange or red coloured
fruits of *R. biflorus,* var. *quinqueflorus, R. innominatus,* and *R.
ichangensis.* At Ichang in early spring the raspberry-like fruit
of *R. parvifolius* is commonly on sale, being locally known as
" Ts'ai-yang-p'ao-tzu." (The term " P'ao-tzu " is comprehensive, covering berries generally.) At Sungpan in August it is
possible to secure fruit of the dwarf *R. xanthocarpus* in quantity
for a few cash pieces.

In the mountains during June and July wild strawberries
are plentiful, and the fruit is of delicious flavour. Two kinds
occur—the white-fruited Hautboy, " Ti-p'ao-tzu " (*Fragaria
elatior*), and the red-fruited " She-p'ao-tzu " (*F. filipendula*).
At Tachienlu, where cream from yak milk is obtainable, I have
enjoyed many a dish of strawberries and cream, and also strawberry pie. By the roadsides the Indian Strawberry (*F. indica*),
also called " She-p'ao-tzu," is everywhere abundant up to
3000 feet altitude. The brightly-coloured but flavourless fruit
of this plant is considered poisonous by the Chinese.

In the woods species of Currant (*Ribes*) with both red and
black fruit are common. One species (*R. longeracemosum*)
bears large, black fruit of good flavour, on racemes 1½ foot
long ! This plant is now in cultivation, and should be
utilized as a parent by the hybridist. A Gooseberry (*R.
alpestre,* var. *giganteum*) is a common hedge-plant throughout
the Chino-Thibetan borderland between 8000 and 11,000 feet
altitude. The small, round, green-coloured fruit is, however,
extremely harsh in flavour. A Strawberry tree, " Yang-mei,"
is common in the margins of woods and thickets between 2000

to 6000 feet altitude, throughout Hupeh, and less so in western Szechuan. The flattened-round red fruit is rough on the exterior, very juicy, and of fair flavour. In the above region the tree so named is *Cornus kousa*. In Yunnan the vernacular name is applied to *C. capitata*, an allied species, but in south-eastern China the " Yang-mei " is *Myrica rubra*, a relative of our Sweet Gale, and belonging to a widely different family.

A climber called " Yang-tao " in Hupeh and " Mao-erh-tao " in Szechuan (*Actinidia chinensis*) is very abundant from 2500 to 6000 feet altitude. It produces excellent fruit of a roundish or oval shape, 1 inch to 2½ inches long, with a thin, brown, often hairy skin covering a luscious green flesh. This is an excellent dessert fruit, and makes a fine preserve. In 1900 I had the pleasure of introducing this fruit to the foreign residents of Ichang, with whom it found immediate favour, and is now known throughout the Yangtsze Valley as the " Ichang Gooseberry." I also was privileged to introduce it into European cultivation, and it fruited in England for the first time in 1911. This valuable climber has, in addition to its edible fruit, ornamental foliage and shoots, and large, fragrant flowers, white fading to buff-yellow. It is a good garden plant ; the only drawback is that the flowers are polygamous, and it is necessary to secure the hermaphrodite form to ensure fruit. Several other species of Actinidia yield edible fruits of fair flavour, one of the best being *A. rubricaulis*, which is now in cultivation.

The Chinese eat the white inner pulp of the pod-like, purple fruits of several species of *Holbœllia* ; these plants known as " Pa-yueh-cha " are stout climbers. The teat-like fruits of several species of *Elæagnus*, known as " Yang-mu-nai-tzu," are also eaten. These have a rather pleasant acid flavour, but are usually astringent in character. The fleshy, thickened fruit-stalks of *Hovenia dulcis*, called " Kuai-tsao," are eaten to annul the effects of wine.

Sweet Chestnut trees are abundant in the woods up to 7500 feet altitude, and excellent nuts, known as " Pan-li," are produced. Several species occur, one of the most common and widely diffused being *Castanea mollissima*. As scrub on the hills up to 3500 feet altitude, the Chinese Chinquapin, " Mao Pan-li " (*C. Seguinii*), is very abundant. Bushes only 2 feet tall produce quantities of small, good-flavoured nuts, but the best are from bushes 5 to 8 feet tall. The best eating chestnuts are, however, those of *C. Vilmoriniana.* This species makes a large tree 60 to 80 feet tall, and has glabrous leaves and a single ovoid nut within each spiny fruit. It is very distinct from all the other members of its family. The acorns of several kinds of Oaks and the nuts of different species of Castanopsis are also eaten by the peasants. This is true of different Hazel-nuts, " Shan-peh-k'o " (*Corylus* spp.), and Beech-nuts (*Fagus* spp.).

A Nut Pine (*Pinus Armandi*) is abundant on the mountains from 3500 to 9000 feet, and the seeds are eaten locally. These seeds, however, are not much sought after, and are far from having the economic importance of the Corean Nut Pine (*P. koraiensis*).

The "Wampi" (*Clausena punctata*) is sparingly grown around Lu Chou.

CHAPTER IV

CHINESE MATERIA MEDICA

NATIVE practitioners in China have very crude ideas of human anatomy, and to be able to " read " the pulse is proof positive of medical skill. Certain foreign drugs like quinine are highly esteemed, but on the whole their faith is in native medicines. Inoculation for smallpox has long been practised, so also has acupuncture for rheumatism, and the value of mercury for certain diseases is well known and it is largely employed. The Chinese materia medica is probably the most varied and comprehensive known. It includes all sorts of the most extraordinary things, ranging from tiger bones to bat's dung, and worse. It is principally, however, vegetable, and the majority of plants found in China are considered to possess medicinal properties to a greater or less extent. Of all this vast array only rhubarb and liquorice have any real value in Occidental practice. The majority of Chinese drugs are supposed to possess tonic and aphrodisiac properties, and the higher a drug is estimated in these respects the greater its commercial value, as witness ginseng and deerhorns in velvet.

The " Father of Chinese medicine " is the Emperor Shen-nung, who, according to legend, ruled from 2737–2697 B.C. This same emperor is also the " God of Agriculture." We are told that Shen-nung went very deeply into the study of herbs, in order to find remedies for the diseases of his people. He is said to have been very successful in his investigations. As an example of his energetic pursuit of this study, it is declared that in one day he discovered 70 poisonous plants and as many that were antidotes to them. Tradition is also responsible for the native belief that he had a glass covering to his stomach, in consequence of which he could watch the process of digestion of each herb and mark its influence on the system. A pharmacopœia, said to have been written by him, formed the nucleus of the " Pun-tsao " or *Herbal*, a great work on the Chinese materia medica. In every druggist's shop of repute there is an image of Shen-nung, and he is looked upon as the presiding deity of the business.

The *Herbal* above referred to was published about A.D. 1590, and its compiler, one Li Shi-chin, spent 30 years in collecting the information. He consulted some 800 previous authors, from whose writings he selected 1518 prescriptions, and added 374 new ones, arranging his materials in 52 chapters in a methodical and (for his day) scientific manner. The work, which is usually bound in 40 octavo volumes, was well received, and attracted the notice of the emperor, who ordered several succeeding editions to be published at the expense of the State. It was, in fact, so great an advance on all previous books, that it checked future writers on the subject, and Li is likely now to be the first and last Chinese critical writer on Natural Science in his mother tongue.

Many curious statements naturally occur in this extensive old work. For example : " The heart of a white horse, or that of a hog, cow, or hen, when dried and rasped into spirit, and so taken, cures forgetfulness." " Above the knees the horse has *night-eyes* (warts), which enable him to go in the night ; they are useful in the toothache." Another is : " If a man be restless and hysterical when he wishes to sleep, and it is requisite to put him to rest, let the ashes of a skull be mingled with water and given him, and let him have a skull for a pillow and it will cure him."

Some very extraordinary remedies are practised to-day. For example : Human milk is supposed to give strength to enfeebled old age. It is considered a meritorious filial act for daughters, granddaughters, and others to thus succour their aged relatives. In Chungking in 1908 an extraordinary case came to my knowledge. A native doctor informed a young woman that the only way to save her mother's life was to administer to her a portion of human liver. This daughter took a large knife and deliberately plunged it into her own body and cut away a portion of her liver. Dr. Asmy, a noble,

self-sacrificing German doctor, working among the Chinese
in Chungking, was informed of the case immediately after it
took place, and succeeded in saving the self-mutilated woman's
life. Dr. Asmy has the piece of liver preserved in spirit and
kept as a memento in his hospital. Among the Chinese soldiers
of the old school it was firmly believed that to eat the heart
of a brave enemy was a sure way of obtaining the courage
he possessed.

These nauseating and nonsensical ideas, however, are not
all taken from the Chinese *Herbal*, and much as we may feel
disposed to smile at the advice contained in this work, it is
well to remember that Western literature on medicine of the
same period contains very much the same sort of instruction.
In Europe as late as the end of the sixteenth century plants
were looked upon from a purely utilitarian point of view,
not only by the masses, but also by very many professed
scholars. Just as men lived in the firm belief that human
destinies depended upon the stars, so they clung to the notion
that everything upon the earth was created for the sake of
mankind ; and, in particular, that in every plant there were
forces lying dormant which, if liberated, would conduce either
to the welfare or injury of man. People imagined they
discerned magic in many plants, and even believed that they
were able to trace in the resemblance of certain leaves, flowers,
and fruits to parts of the human body, an indication emanating
from supernatural powers, of the manner in which the organ
in question was intended to affect the human constitution.
The similarity in shape between a particular leaf and
the liver did duty for a sign that the leaf was capable of
successful application in cases of hepatic disease, and the
fact of a blossom being heart-shaped must mean that it would
cure cardiac complaints. Thus arose the so-called Doctrine
of Signatures, which, brought to its highest development by
the Swiss alchemist, Bombastus Paracelsus (1493-1541),
played a great part in the sixteenth and seventeenth centuries,
and still survives at the present day in the mania for nostrums.

In ancient Greece there was a special guild, the " Rhizotomoi," whose members collected and prepared such roots
and herbs as were considered to be curative, and either sold

A LOAD OF "TU-CHUNG"
(THE BARK OF EUCOMMIA ULMOIDES)

A MEDICINAL RHUBARB (RHEUM OFFICINALE)

them themselves or caused them to be sold by apothecaries. The " Medicine Guild " in China to-day performs much the same work, and its origin is long anterior to the Greek Rhizotomoi. If, then, Chinese pharmacology is to-day several centuries behind that of the Occident, there was a time when it was equally far in advance. Marco Polo makes many references to the value of Chinese drugs. For example : " All over the mountains of the province of Tangut, rhubarb is found in great abundance, and thither merchants come to buy it, and carry it thence all over the world."

All parts of the Chinese Empire contribute something to the native pharmacy, but, with the exception of ginseng, cassia-bark, camphor, and areca-nut, nearly all the more highly valued drugs come from the forests and scrub-clad highlands of the west. The famous drug, ginseng, the root of *Aralia quinquefolia*, comes from Corea and Manchuria, and the best quality sells for its weight in gold. To the Chinese this drug is the *radix vitæ*, restoring strength, vitality, and power to old and young. So precious is this " life-giving root " that the best plants are, in theory, reserved entirely for the emperor's use. On the Chinese system this drug unquestionably acts as a strong restorative, tonic, and aphrodisiac, adverse Western opinion notwithstanding. In the forests of the west certain " bastard ginsengs " occur, but are little valued.

Cassia-lignea, the bark, buds, and leaves of *Cinnamomum Cassia*, comes from certain districts (Luk-po, Lo-ting) in Kwangtung and (Tai-wu) in Kwangsi, provinces in the south, where it is largely cultivated and exported to all parts of the Empire and elsewhere. Cassia-bark, " Kuei-p'i," is valued as a tonic, stimulant, and condiment. Areca-nut, the seed of a palm (*Areca catechu*), occurs in these southern provinces, and also in Yunnan. It is also imported from Cochin-China. Betel-chewing is not in general vogue among the Chinese, who value the nuts more as a medicine, chiefly as an astringent and anthelmintic.

Camphor is in general use all over China. The most valued kind is the Baros Camphor (*Dryobalanops Camphora*), imported from Malaysia (Borneo), the camphor produced in Japan, Formosa, and Fokien from *Cinnamomum Camphora* being less esteemed, and chiefly valued for purposes of export to other parts of the world. The Chinese value Baros Camphor as a tonic and aphrodisiac.

The Imperial Maritime Customs officials have paid considerable attention to Chinese medicines, and in 1889 a list was published by order of the Inspector-General, the late Sir Robert Hart. This list was compiled from the Returns of each Treaty port, and an attempt was made therein to identify the plants yielding the drugs and to give their province of origin. The difficulties besetting such a task were enormous, but much good work was accomplished. Consul-General Hosie, in his *Report on the Province of Ssuch'uan*, has compiled a list of Szechuan medicines which very accurately represents the present state of our knowledge on this subject. Not until a complete collection of herbarium material covering flowers and fruits is made, and the whole submitted for identification to some one or other of the great herbaria in Europe, will it be possible to assign correct scientific names to a vast number of these medicines.

Hosie's list comprises 220 different kinds, of which number 189 are of vegetable origin. The trade importance of drugs is enormous; the exports passing through the Maritime Customs, at the port of Chungking, in 1910, being valued at over Tls. 1,540,000; those from Hankow at over Tls. 1,780,000.

I do not propose entering into a detailed account of the Chinese medicines, but will briefly note a few of the more important and their uses, which may not be without interest. Perhaps the most generally useful drug known from China is rhubarb, " Ta-huang." The Rhubarb plant occurs throughout the highlands of the Chino-Thibetan borderland, but, as in the days of Marco Polo, the best comes from the " Province of Tangut." This region stretches from Sungpan in a northwesterly direction, and includes part of the modern province of Kansu. Rhubarb is found growing among scrub and near rocky watercourses between 7500 feet and 12,500 feet altitude. It is also commonly cultivated, but the wildling is esteemed the best drug. The finest rhubarb is obtained from the plant known botanically as *Rheum palmatum*, var. *tanguticum*, and

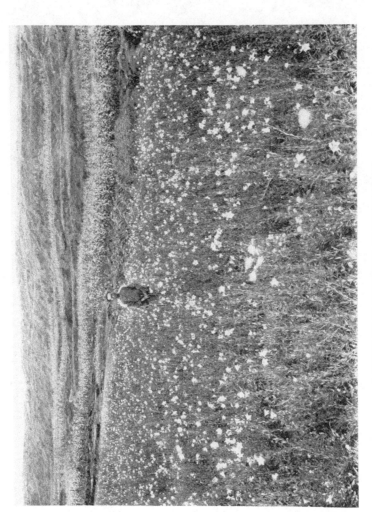

A FIELD OF PLATYCODON GRANDIFLORUM

CHINESE MATERIA MEDICA 39

this is the variety most commonly met with throughout the extreme north-west of China and the contiguous Thibetan regions. From Tachienlu are exported considerable quantities of a second-grade rhubarb, which is mainly derived from *R. officinale*, although the variety *tanguticum* also occurs sparingly in that neighbourhood. Other species of Rheum grow in the west, and are used as adulterants. In north-western Hupeh *R. officinale* occurs in the forests, and is also cultivated by the peasants, but the quality of the drug is very poor. The so-called " Tangut regions " enjoy a dry, sunny climate, and curing the drug is a much easier task than in the other districts mentioned. This also probably affects the quality. In China, rhubarb is valued as a purgative; and is employed in the same way as in the Occident.

The best liquorice, " Kan-tsao " (Sweet-herb), is also a product of the grasslands north-west of Sungpan ; inferior kinds grow elsewhere in China. The source of the Sungpan product has been identified as *Glycyrrhiza uralensis*. It is valued as an emollient, and small quantities enter into nearly every prescription intended for internal application. The drug known in the vernacular as " Ch'ung-tsao " is a caterpillar infested with the mycelium and the projecting fructification of a Fungus (*Cordyceps sinensis*). This is another valued product of the western uplands, where it is found at from 12,000 to 15,000 feet altitude. The body of the caterpillar is yellowish, the fructification of the fungus black, the two together being sticklike in appearance and about 5 inches in length. As a medicine it is esteemed for a variety of purposes—boiled with pork it is employed as an antidote for opium-poisoning and as a cure for opium-eating ; also with pork and chicken it is taken as a tonic and mild stimulant by convalescent persons, and rapidly restores them to health and strength.

The tiny white bulbs of the Fritillary (*Fritillaria Roylei;* and other allied species), known as " Pei-mu " or " Jên Pei-mu," constitute one of the most highly valued medicines from the alpine regions of the west, where the plants grow at from 12,000 to 15,000 feet altitude. Large quantities of this drug are exported from Monkong Ting and Tachienlu. The bulbs are pounded, then boiled with dried orange skin and sugar. The resultant is taken as a cure for tuberculosis and asthma.   In Hupeh the pseudo-bulbs of two terrestrial Orchids, *Pleione pogonioides* and *P. Henryi*, are used for the same purpose, and are known as " Ch'uan Pei-mu." These plants grow on moist, humus-clad rocks in the woods between 3000 and 5000 feet altitude.

In clearings in the woods throughout western Hupeh and on Mount Omei plantations of Huang-lien (*Coptis chinensis*) are maintained as a profitable investment.   The dried rhizome is an all-round medicine, and particularly valued as a stomachic. An infusion is considered a cure for dyspepsia ; used by women nursing children, it is said to promote the flow of milk ; pounded and mixed with the white of eggs it is applied as a poultice to boils.   Personally I can testify that it makes an excellent and appetizing bitters.

The thickened roots of a number of umbelliferous plants are esteemed for their medicinal virtues, as tonics and blood puri- fiers generally.   One in general use and commonly cultivated is " Tang-kuei " (*Angelica polymorpha*, var. *sinensis*).   An ex- tract obtained by boiling the root-stock of *Platycodon grandi- florum*, a campanulaceous plant known locally as " Chieh- k'eng," is a cure for chill in the stomach.   The small pods of *Gleditsia officinalis*, " Ya-tsao," sliced and boiled with " Tang- kuei," forms an infusion which is considered a certain cure for coughs and colds.

For medicinal purposes the Aconite, " Tsao-wu-tu " (*Aconi- tum Wilsonii*), is cultivated, the powdered root being mixed with the white of eggs and applied externally as a remedy for boils.   The " Ch'uan-wu-tu " (*A. Hemsleyanum*, and other climbing species) has similar uses to the foregoing.   Also after frequent boilings the root is used in minute quantities as a drastic cure for coughs.   Another twining herb, " Tang-shên " (*Codonopsis tangshen*), is commonly cultivated in the moun- tains, the thickened root-stock being valued as an all-round tonic.

The barks of many trees are used in medicine, and the identification of these is not so difficult as in the case of the herbs. One of the most esteemed is " Hou-p'o " (*Magnolia officinalis*). The best quality bark is worth 1000 cash per ounce.   An extract is taken as a tonic, aphrodisiac, and a certain cure for colds, all in one. The dried flower-buds of this tree called " Yu-p'o " yield an extract by boiling which is taken by women to correct irregularities of menstruation. The bark of *Eucommia ulmoides*, " Tu-chung " or " Tsze-mien," is pounded and boiled, the extract being taken with wine and pork as a cure for troubles of the kidney, liver, and spleen. It is also supposed to be a diuretic and aphrodisiac, and is a valuable general tonic. The bark of *Picrasma quassioides*, " Ku-lien-tzu," yields, on boiling, an extract which is used in cases of colic and pains in the stomach generally ; also as a febrifuge. The bark of *Phellodendron chinense*, " Huang-po " or " Huang-peh," is a complete materia medica in itself, it being used internally and externally as a general remedy for almost every ailment known to the Chinese, and, being cheap, is a poor man's " cure-all."

These selected examples, although few in number, are perhaps sufficient for the purpose of this chapter. Undoubtedly many of the drugs used by the Chinese possess sound medicinal properties, and their proper investigation is well worth the attention of Occidental pharmaceutists.

CHAPTER V

GARDENS AND GARDENING

FAVOURITE FLOWERS CULTIVATED BY THE CHINESE

ORNAMENTAL gardening has been practised in China from time immemorial, and the people are endowed with an innate love for flowers and gardens. Floral calendars are kept in every house above the poorest, and volumes of poems have been written in praise of the Moutan Pæony, Camellia, Plum, Chrysanthemum, Lotus-lily, Bamboo, and other flowers. The appearance of the blooms on the more conspicuous flowering shrubs is eagerly watched for, and excursions into the country are taken to enjoy the sight of the first bursting into blossom of favourite plants. The dwelling of the poorest peasant is usually enlivened by an odd plant or two, and the courtyard of the shopkeeper and innkeeper always boasts a few flowers of one sort or another. The temple grounds are frequently very beautiful, and attached to the houses of the cultured and wealthy are gardens often of great interest. In the neighbourhood of wealthy cities like Soochou, Hanchou, and Canton, are public and private gardens which are famed throughout the length and breadth of the Empire. The finest example I have seen is fittingly associated with the emperor's summer palace, a few miles outside Peking. There Chinese gardening may be seen at its best, and it calls forth admiration from all visitors.

Chinese landscape-gardening is represented at its best in the so-called " Japanese gardens " of to-day. The Japanese have undoubtedly carried the art to a higher state of perfection than the Chinese, but the latter unquestionably originated it. In all these gardens the love of the grotesque predominates, and the landscape effect is essentially artificial; yet in accord-

A CYPRESS AVENUE (CUPRESSUS FUNEBRIS), CHAO-CHUEH TEMPLE NEAR
THE CITY OF CHENGTU

ance with their own ideals the Chinese are most skilful and accomplished gardeners. Given a piece of ground, no matter if it be small, and devoid of all natural beauty, or badly situated, they will patiently transform it into a mountain-landscape in miniature. With strange-looking, weather-worn rocks, dwarfed trees, bamboos, herbs, and water, a piece of wild country-side is evolved replete with mountain and stream, forest and field, plateau and lake, grotto and dell. A network of narrow winding paths traverses the garden, and rustic bridges in various designs are thrown across the infantine streams. The whole effect is often encompassed within a comparatively few square yards, though the perspective is one of seemingly many miles. In all the larger gardens, closely associated with and usually in part overhanging a pool where the Lotus-lily is grown, a small pavilion is erected. Here the proprietor and his guests resort to drink tea or wine, chat, and admire the various flowers. When no male guests are present the garden is frequented by the female members of the family, with whom it is ever a favourite sanctum.

The Chinese do not cultivate a very great variety of plants, and the contents of the various gardens are much the same, though necessarily the selection is modified by climate and locality. To all the flowers grown in Chinese gardens some peculiar significance or æsthetic value is attached. An orchid (*Cymbidium ensifolium*), called "Lan hwa," is regarded as the "king of flowers," the modest appearance of the plant. and the delicate odour of its blossoms, representing the very essence of refinement. The "Mei hwa" (*Prunus mume*), owing to the beauty and perfume of its flowers, which are produced in winter when few plants are in blossom, is very highly prized and regarded as a "flower of refinement." Around Peking the same vernacular name and attributes are attached to *P. triloba* and its double-flowered form. The Winter-sweet, "La-mei hwa" (*Meratia præcox*), is similarly esteemed.

The various Bamboos, emblems of grace and culture, and beautiful at all seasons of the year, are indispensable garden plants. "No man can live without a Bamboo tree in the immediate vicinity of his house, but he can live without meat," is a favourite Taouist saying. The Chrysanthemum, "Chu hwa," and Moutan Pæony are other " flowers of refinement " almost reverently appreciated for the colour and beautiful form of their flowers. The Lotus-lily, " Lien hwa " (*Nelumbium speciosum*), is regarded as an emblem of purity, and the Goddess of Mercy (Kwanyin) is always represented seated in the centre of a Lotus flower. The Chinese " Luck Lily " or " Water Fairy " (*Narcissus tazetta*) is cultivated in vast quantities, more especially throughout the eastern part of the Empire, and is in blossom for the New Year festival. It is appreciated for its odoriferous flowers, and its luxurious growth is considered prophetic of wealth and prosperity. This Narcissus is not a Chinese plant, but is a native of the Mediterranean region, from whence it was long ago introduced into China by Portuguese traders, and it together with the Pomegranate are virtually the only exotic flowers in high favour with the Chinese.

The Pearl Orchid, " Chu-lan hwa " (*Chloranthus inconspicuus*), is valued for the delicate odour of its flowers, which are used in the Anhui province in scenting green tea for the Chinese market. Table grass (*Liriope spicata*) is admired for its graceful habit, and is placed on a desk or table, to afford rest to the eyes when reading or studying. Lastly in this relation may be mentioned the " Hoary Pine," which is emblematical of revered old age. This name is applied to several kinds of Conifers other than Pinus proper.

To complete the list of favourite Chinese flowers we may enumerate Camellia, Heavenly Bamboo, " Tien-ch'u " (*Nandina domestica*), " Kuei hwa " (*Osmanthus fragrans*), " Tzu-ching " (*Lagerstrœmia indica*), " Tiao-chung " (*Enkianthus quinqueflorus*), " Chin-yin hwa " (*Lonicera japonica*), numerous varieties of Azaleas, Roses, Balsams, Cockscombs, doubleflowered Peaches, various Conifers, and Box (*Buxus japonica*). Some or all of the above are to be found in every Chinese garden of note. Though the cultural skill expended on many of them is in the direction of dwarfing and training into grotesque shapes, this treatment in no sense robs the flowers of the qualities attributed to them in literature and song. The decoration found on Chinese porcelain well illustrates the nation's love of beautiful flowers and quaintshaped trees.

THE LOTUS LILY (NELUMBIUM SPECIOSUM)

China is a land of contrariety—a land whereof no general statement or observation holds good. In spite of their love for the grotesque and the artificial landscapes seen in their gardens, the Chinese have a strong appreciation of natural beauty. This is evidenced by the sites chosen for their temples and shrines and for the tombs of the wealthy. Apart from situation, which is usually perfect, such sanctuaries always nestle beneath the shade of magnificent trees, and are approached as a rule through avenues or groves of large trees. Though a few deciduous trees are commonly found, evergreens always have distinctive preference. In the temple grounds around Peking are noble avenues of Arbor-vitæ (*Thuya orientalis*), Juniper (*Juniperus chinensis*), Elm (*Ulmus pumila*), and Sophora (*S. japonica*) ; in the south, centre, and west of the Empire, Pine (*Pinus Massoniana*), China Fir (*Cunninghamia lanceolata*), Cypress (*Cupressus funebris*), Nanmu (*Machilus nanmu*, and allied species), "Yu-la shu" (*Photinia Davidsoniæ*), Wintergreen (*Xylosma racemosum*), Banyan (*Ficus infectoria*), and a few other kinds of trees are always present. Many of these trees are extremely rare, except in the precincts of religious sanctuaries.

The world at large does not realize how deeply it is indebted to religious communities for the preservation of many trees. In Europe, for example, most of the best varieties of Pears originated in the gardens attached to religious establishments in France and Belgium and were introduced into England and other countries after the battle of Waterloo. In China, where every available bit of land is devoted to agriculture, quite a number of trees must long ago have become extinct but for the timely intervention of Buddhist and Taouist priests. The most noteworthy example of this benevolent preservation is the Maidenhair tree (*Ginkgo biloba*). This strikingly beautiful tree is associated with temples, shrines, courtyards of palaces, and mansions of the wealthy throughout the length and breadth of China, and also in parts of Japan. But it is nowhere truly wild, and is a relic of a very ancient flora. Geological evidence shows that it is the last survivor of an ancient family, which flourished during Secondary times, and can even be traced back to the Primary rocks. In Mesozoic

times this genus played an important part in the arborescent
flora of north-temperate regions. Fossil remains, almost
identical with the present existing species, have been found,
not only in this country and North America, but also in
Greenland. Though to-day Chinese gardens, nurseries, and temple
grounds do not contain anything new in the way of ornamental
or economic plants, it was otherwise up to the middle of the
last century. Our early knowledge of the Chinese flora was
based on plants procured from gardens, notably from those
around Canton.   The plants were brought to Europe by trading
vessels, especially those of the East India Company, at the
end of the eighteenth and early in the nineteenth centuries.
Different patrons of horticultural and botanical institutions
in England lent financial assistance, and collectors were dis-
patched to investigate and send home all that they could
possibly find.

By these means our gardens first secured the early varieties
of Roses, Camellias, Azaleas, greenhouse Primroses, Gardenias,
Moutan Pæonies, Chrysanthemums, Chinese Asters, and such-
like familiar plants. The Chrysanthemum, for instance, has
been cultivated in China and Japan from time immemorial,
and its parent forms (*Chrysanthemum sinense* and *C. indicum*)
are common wild flowers around Ichang and elsewhere in
China. In Europe *C. sinense* was first cultivated in Dutch
gardens as early as 1689, no less than six kinds being then
known. But these were subsequently lost, and when the
plant was again introduced in 1789, through the agency of
Sir Joseph Banks, the plant was unknown to Dutch gardeners.
The famous gardener, Philip Miller, cultivated *C. indicum* in
the Chelsea Physic Gardens in 1764, it having been discovered
in 1751, near Macao, south China, by Osbeck. This species
has, however, had much less to do in the evolution of our
present-day Chrysanthemum than *C. sinense*.

The parent of our Tea Roses is *Rosa indica*, the Chinese
Monthly Rose, long cultivated in China and still to be found
wild in the central and western parts of the Empire. It was
introduced into England through the efforts of Sir Joseph
Banks in 1789. The parent of our greenhouse Primroses

THE MAIDENHAIR TREE (GINKGO BILOBA) 90 FT. TALL, GIRTH 24 FT.

(*Primula sinensis*) was introduced from Canton into the garden of Thomas Palmer, Esq., of Bromley, Kent, by John Reeves about 1820. It is a native of Hupeh, where it occurs in great abundance on the dry, precipitous, limestone cliffs of the Ichang Gorge and its lateral glens. The wildling is a true perennial with flowers of a uniform mauve-pink colour. Another greenhouse Primrose (*P. obconica*) occurs in the same region but in moist loamy situations.

The Indian and Mollis Azaleas and a score of other favourite plants of our gardens all came originally from Chinese gardens through various agencies. It is true we have developed most of these introductions almost beyond recognition, and the Chinese are now acquiring new forms and varieties from us, yet without these early arrivals how much poorer our gardens and conservatories would be to-day! In bygone times, even only about a century ago, that part of the world which we know as China was loosely spoken of as the " Indies," and this geographical blunder is perpetuated in the specific name " indica " which botanists have attached to some of these plants. In the middle of last century many ornamental plants were received from the gardens of Japan, and botanists, assuming that these were natives of the country, gave the specific name "japonica" to certain of them. Subsequent knowledge has, however, conclusively proved that a number of the so-called Japanese plants are only cultivated forms of plants originally natives of China. Thus has the geographer and botanist unwittingly obscured China's right to be termed the " Kingdom of Flowers."

CHAPTER VI

AGRICULTURE

THE PRINCIPAL FOOD-STUFF CROPS

THE Chinese might appropriately be termed a " Nation of Shopkeepers," yet, in spite of their commercial enterprise, the agricultural industry is the backbone of the nation. With a vast population to support, every possible inch of land has been brought under cultivation, and prodigious efforts have been made to obtain the greatest returns from the soil. In spite of it all, millions are ever on the verge of starvation, and almost annually either drought or flood brings famine to some part of the Empire.

Landed property is held in clans or families as much as possible and is not entailed, nor are overgrown estates frequent. The land is all held directly from the Crown, no freehold being acknowledged. The conditions of common tenure are the payment of an annual tax, the fee for alienation, with a money composition for personal service to the Government. The proprietors of land record their names in the district and take out an original deed (called " red deed ") which secures them in possession as long as the ground tax is paid. This sum varies very much according to the fertility, location, and nature of the land, but is nowhere heavy or severe. Naturally, good rice-land pays the heaviest tax. The paternal estate, and the property thereon, descends to the eldest son, but his brothers can remain upon it with their families and devise their portion *in perpetuo* to their children, or an amicable composition may be made ; daughters never inherit, nor can an adopted son of another clan succeed. A mortgagee must enter into possession of property and make himself responsible for the payment of taxes levied thereon. The enclosure of recent alluvial

48

THE RAMBLER ROSE (ROSA MULTIFLORA)

deposits cannot be made without the cognizance of the authorities, but the terms in such cases are not onerous. When waste hillsides and poor areas are brought under cultivation ample time is allowed for a return of the capital expended in reclaiming them before assessment is made.

Since the food-supply of the Chinese population has always been supplied from within the Empire, agriculture has rightly been accorded first place among all branches of labour from time immemorial. According to legend, the Emperor Shennung (2737–2697 B.C.) established agriculture as a science. He examined the various kinds of soils and gave directions as to what should be cultivated in each. He taught the people how to make ploughs, and instructed them in the best methods of husbandry. Immediate results were seen in the improved conditions of the people, and succeeding generations have amply testified their gratitude to him. Under the title of " Prince of Cereals " he has long since been deified, and is worshipped throughout the length and breadth of the land. In Peking there is an altar dedicated to him, enclosed within a large park. Formerly, at the vernal equinox, the ruling emperor, assisted by various officials, performed an annual commemorative ceremony of ploughing a portion of the park.

The Chinese nation is to a very large extent vegetarian, flesh being eaten only in small quantities except on festival occasions. Pork, chickens, ducks, and fish comprise the meat-diet, and of these the Chinese are excessively fond, but to the great majority they are luxuries, only to be indulged in on rare occasions. Rice is to them what wheat is to us, only more so. So long as the average Chinaman can get rice he is happy ; but this would be scarcely true of ourselves if we could only get bread ! Next to rice the more important food-stuffs are wheat, maize, pulse, and cabbage. The Chinese fry most of their vegetables, and for this purpose a vegetable oil is nearly always used. The oils expressed from the seeds of members of the Cabbage (*Brassica*) family, the Soy Bean (*Glycine hispida*), and Sesamé (*Sesamum indicum*) being most in request.

Whilst the Chinese cultivate a great variety of vegetables the quality of one and all, judged from our standard, is wretchedly inferior. With the exception of maize and sweet

potatoes, it is safe to say that not a single Chinese vegetable
would command attention in this country. In this chapter
I have attempted a fairly exhaustive account of this subject, in
so far as it came under my observation during the eleven years
I travelled in China. These observations were mainly limited
to three provinces, namely, Yunnan, Hupeh, and Szechuan.
The estimated area of these territories is about 372,500
square miles—more than three times that of the United
Kingdom. Other parts of China have vegetables peculiarly
their own. Again, at the treaty ports, where foreigners have
settled, varieties of our own vegetables have been introduced
and are cultivated for their use. These, with rare exceptions,
do not come within our province.

In China the fields are all so small that market-gardening
rather than farming best describes the agricultural industry.
Long experience has taught the people how to obtain the
maximum returns without unduly exhausting the soil, indeed,
the extraordinary thing about Chinese agriculture is the fact
that, although cultivation has been so long in progress, the soil
shows practically no sign of exhaustion. Artificial manures are
unknown to the Chinese farmer, and ordinary farmyard manure
is scarce and almost a negligible factor. Constant tillage, aided
by as much sewage as can be possibly obtained, are relied upon
to produce full crops. The sewage from cities and villages is
carried long distances away in buckets or in tubs to the fields,
and nowhere else in the world is human excrement so highly
valued or so laboriously collected. In matters of seed-selection,
plant-breeding, and the higher arts of agriculture, the Chinese
have everything to learn. Rotation of crops and the enrich-
ment of the soil by leguminous crops they understand and
practise as fully as circumstances permit.

Rice (*Oryza sativa*) is, of course, the favourite cereal, but being
a tropical plant, requiring an aquatic habitat, its area of cultivation is restricted in China, and probably a third of the
people never taste this grain save on festival occasions. In
southern China two crops of rice are obtainable annually, but
throughout the greater part of the land where this cereal is
cultivated only one crop can be grown in a season. This
occupies the ground from May until early in September.

In the cultivation of rice, the patience, ingenuity, and incredible industry of the Chinese are particularly well exemplified. The terraced fields, necessary to ensure a flow of water, whether it be on a seemingly flat plain or on a steep hillside, meet the eye of the traveller on all sides. It is little short of marvellous when one reflects on the skilful way in which the entire rice-belt in China is terraced, and the enormous amount of time and labour involved in the undertaking indicate what a hard task-master necessity has been. In matters of irrigation the Chinese are past masters. They have not yet succeeded in making water run uphill, but with their various contrivances they lift it bodily from streams and ditches and convey it long distances to wherever it is needed. The number of devices for irrigation purposes is almost legion, and though simple in principle and efficacious in results they are intricate in detail. Some are operated by hand, others by the foot, and many are automatically worked by the current of the streams. The large skeleton-like water-wheels depicted in the photographic illustration (p. 52), represent one of the methods commonly in use in central and Western China.

Rice-cultivation presents many tedious details and the layman will probably find it difficult to realize that in China the whole crop is planted by hand. The grain is sown thickly in nursery-beds, and when the seedlings are 5 or 6 inches tall they are transplanted in small clumps equidistant in the flooded, prepared fields. Men and women take part in this work, and it is surprising how rapidly the fields are planted. The rice plants are made firm in the mud by treading around them immediately they are established. The fields are kept free of weeds and the requisite supply of water is maintained until, as the crop ripens, the fields are finally allowed to get dry. The rice crop is reaped by hand, and without being removed from the field the grain is at once beaten off into wooden bins ; afterwards it is dried and stored. The Chinese cultivate three well-marked varieties of rice—namely, ordinary, red, and glutinose. The first two are grown for food only; the red, being the hardiest, is cultivated at higher altitudes than the other, but is by no means confined thereto. This Red Rice, " Hung-mê " (*O. sativa*, var. *præcox*), gets its name from the reddish colour of the pellicle, which adheres tenaciously to parts of the grain after milling. Glutinose Rice (*O. sativa*, var. *glutinosa*) does not take the place of the other two as a food-stuff, being only eaten for a change. It is valued for the weak spirit which is made from it, for the sugar which is extracted from it, and for making into cakes and sweetmeats. It is later in ripening than the other varieties, and always commands a higher price in the market. In Yunnan a variety which will thrive without water is grown. This Upland Rice (*O. sativa*, var. *montana*) yields but a poor crop and is very inferior.

Whilst the Chinese are pre-eminently a rice-eating race, it should be borne in mind that there are millions of Chinese who, save on rare occasions, never eat rice at all. To these people, wheat, maize, and buckwheat are the staple cereals. In the rice-growing districts of China, Wheat (*Triticum sativum*) is a winter crop, occupying the ground from October to early May. In the mountainous districts and in the colder provinces it is a most important summer crop. I have noted no fewer than five very distinct varieties, comprising both " red " and " white " wheats, and both awned and awnless kinds. In late August the mountain-sides and valleys in western Szechuan present a glorious picture of miles and miles of rolling grain fields. In this region 8000 to 10,500 feet represent the wheat-growing belt. The grain is sown by hand in rows, the seeds being dropped in clusters a few inches apart. In the Yangtsze Valley, if the wheat crop is late in ripening, it is ploughed in to make way for rice. In the plains of central China the grain is threshed out the moment it is harvested. On the Thibetan borderland it is tied into sheaves and stacked, ears downwards, on tall hurdle-like arrangements (Kai-kos) until time and weather admit of its being threshed. (These remarks also apply to barley, oats, and other crops.) The grain is ground into flour and made into cakes and vermicelli. Chinese flour is usually gritty and of bad colour.

Barley is sparsely cultivated throughout the Yangtsze Valley, and it is only in the mountainous Thibetan borderland that it is largely grown. The Chinese do not care for the meal, and the grain is chiefly used for making spirit and for feeding pigs and other domestic animals. The Thibetans, on the other

IRRIGATION WHEELS

hand, highly esteem barley. Roasted and ground into meal and mixed with tea and rancid butter it forms " Tsamba," their national and staple food. Since it is hardier than wheat its culture extends to a greater altitude; the highest point at which I noted it was 12,000 feet. Both Chinese and Thibetans cultivate several varieties, but the six-rowed variety, *Hordeum hexastichon*, is most in favour. Around Sungpan a variety of the above, having purple paleæ, is largely grown, being considered hardier than the type. This variety is apparently peculiar to this region, being quite distinct from the two-ranked chocolate Barley (*H. cœleste*), which is cultivated in parts of the Himalaya. Ordinary Barley (*H. vulgare*) is cultivated in smaller quantities than the preceding kinds by Chinese and Thibetans. In Hupeh and in the river-valleys of western Szechuan I met with occasional patches of *H. hexastichon*, var. *trifurcatum*. This variety is the Mi-mê (Rice-wheat) of the Chinese.

In the mountains Rye (*Secale fragile*) is sparingly grown and the grain eaten.

Oats are not much grown by the Chinese in the parts through which I travelled, but they are cultivated to a considerable extent by the Thibetan and other tribesmen in the highlands. The Chinese prefer *Avena nuda*, which they designate " Yen-mê " ; the Thibetans and tribesfolk favour *Avena fatua*. The grain of both these kinds is roasted and ground into oatmeal, or cooked and eaten whole.

Next to rice and wheat, Maize, or " Pao-k'o " (*Zea Mays*), is the most important cereal. This plant is of American origin, but it has been so long cultivated in China that the date of introduction is not ascertainable. In the rice-belt it is relegated to land that for one reason or another is not suitable for rice. It is in the more mountainous parts that maize is the staple crop. It occupies the gullies and slopes of the mountains, and commonly so steep are these that one wonders how the people manage to sow and reap the crop. Wild pigs rob the maize fields, and when the crop is in ear the farmers beat gongs and make as much noise as possible during the night to scare these animals away. In open country tall thatched look outs are erected, where the juvenile and female members of the family sit and watch for thieves during the daylight.

In the Yangtsze Valley maize is always a summer crop, and two crops are frequently harvested. In the mountains its cultivation extends up to 8000 feet, and in exceptionally favourable districts even higher. Green corn is really a delicious vegetable, and ought to be used in this country. The Chinese, however, do not employ it extensively in this form. When ripe the sheaths of the cobs are folded back, exposing the grain. They are then tied in bunches and suspended from the roofs of houses, where they can be kept dry. The grain is ground and made into meal-cakes ; it is also used for making spirit. From the culms sugar is sometimes extracted, but their chief use is for fuel.

False Millet (*Sorghum vulgare*), the " Kao-liang " or " Hsu-tzu " of the Chinese, is largely used for making wine. It is cultivated generally throughout central and Western China, but not so extensively as in the northern parts of China, and notably Manchuria. The largest areas I noted were on the plateaux of Yunnan, the plain of Chengtu, and the fluviatile areas of the Min and Fou Rivers. Its altitudinal limit is about the same as that of maize, and, like this latter, it is always a summer crop. Two distinct varieties are grown, one with purple, the other with yellowish " heads." It is occasionally employed as food, more particularly in mountainous districts, but 90 per cent. of it is used for making wine.

Other Millets met with in cultivation are *Panicum miliaceum*, " Chan-tzu " ; *Setaria italica*, " Hsiu-ku " ; and *Panicum crusgalli*, var. *frumentaceum*, "Lung-tsao-ku," but not in large quantities. The grain is used in making cakes and for feeding bird-pets. The cereal commonly known as Job's Tears, " Ta-wan-tzu" (*Coix lachryma*), is cultivated in small patches throughout central and Western China. Though occasionally used as food in the form of gruel, " Job's tears " are chiefly valued as medicine. They are supposed to possess tonic and diuretic properties, and are administered in cases of phthisis and dropsy.

Of Buckwheat two species are commonly cultivated, namely, *Fagopyrum esculentum* and *F. tataricum*, the "T'ien-ch'iao-mê " and " K'u-ch'iao-mê " respectively of the Chinese. These constitute a most important crop, especially in the highlands. Under favourable climatic conditions two crops are harvested.

A field of the pink Buckwheat (F. *esculentum*) in flower is one of the prettiest sights imaginable. It is most commonly grown on terraced mountain-sides. The other species grows twice as tall as the above, and bears greenish-white flowers. The altitudinal limit of buckwheat equals and possibly exceeds that of barley. After the seeds are threshed out they are ground up in water, and the husks are removed by a fine sieve. The flour is then made into dough with a little salt, to which lime is added. This dough is made into vermicelli, when it is ready for cooking and eating. Buckwheat constitutes a most important article of food among the Chinese who live in the mountainous districts, and also with the tribesfolk of the borderland. It is a very accommodating crop, for it thrives on the poorest of soils, requires little attention beyond sowing and harvesting, and matures very quickly.

Since the Chinese are to such a large extent a vegetarian people, the various members of the pea and bean family are necessarily most important crops. The Common Pea, " Mê-wan-tzu " (*Pisum sativum*), and Broad Bean (*Vicia Faba*), with the Soy Bean (*Glycine hispida*), are the most important. The two former are winter crops in the valleys and summer crops in the highlands. The soy bean is everywhere a summer crop. Peas and broad beans are eaten both fresh and dried. They are also ground into flour and made into vermicelli. The young shoots of the pea are eaten as a vegetable. The soy bean, " Huang-tou," is of even greater value than the preceding ; it is planted everywhere—in fields by itself, around rice and other fields, and as an undercrop to maize and sorghum. It yields seeds of three colours, namely, yellow, green, and black. The Chinese distinguish three kinds of the yellow and two kinds each of the green and black. These varieties yield a succession of beans, the black being fully a month later than the others. The " Huang-tou " is cooked and eaten as a vegetable, or ground into flour and made into vermicelli ; preserved in salt it makes an excellent pickle. It is also extensively used in the manu-facture of soy sauce and soy vinegar. A variety with small yellow seeds is largely employed in making bean-curd. While in central and Western China the soy bean is cultivated ex-clusively as a food-stuff, in Manchuria it is grown almost solely for the oil which is obtained from the seeds by pressure, and for the residual-cakes that remain after the oil had been expressed. From Newchwang, the port of Manchuria, there is an enormous export trade done in " Bean-cake," which is in great demand as an agricultural fertilizer in all parts of China. The soy bean has recently been exported to Europe in large quantities, and the soy-bean oil is employed in soap-making and for culinary purposes.

Two kinds of Gram, *Phaseolus mungo,* " Lu-tou," and *P. mungo,* var. *radiatus,* " Hung-tou," are grown as summer crops. The seeds of the " Lu-tou " (green bean) are especially valued for their sprouts. To obtain these the beans are put in jars with water and covered over. Under these conditions they quickly develop shoots a couple of inches or more long, which are highly esteemed as a vegetable. Of the " Hung-tou " (red bean) there are two or three varieties. The seeds of these are used as a vegetable or ground into flour and employed for stuffing cakes and sweetmeats.

The Lentil (*Ervum Lens*), " Chin-mê-wan-tzu," is cultivated as a winter crop, being commonly associated with peas and broad beans. It is, however, by no means extensively grown. The seeds are eaten cooked. Oil is occasionally expressed from the seeds and used for lighting purposes. Other pulses are *Dolichos Lablab,* " Pien-tou," of which there are several varieties, *Canavalia ensiformis* (Sword Bean), *Phaseolus vulgaris,* " Yün-tou," *Vigna Catiang,* and *Cajanus indicus,* all commonly and extensively cultivated. Though the seeds of the first four are eaten, it is more for the pods, which are sliced, cooked, and eaten as a vegetable, that these plants are valued. The cylindrical pods of *Vigna Catiang* are from 1½ to 2 feet long, and about the thickness of a lead-pencil. Though the Chinese esteem it, I have found it only a very tasteless vegetable. As a winter crop in parts of the Yangtsze Valley, *Melilotus macrorhiza,* " Yeh-hua-tsen," is sparingly cultivated. The green shoots are sometimes eaten as a vegetable ; the seeds are used medicinally for colds.

Of cabbages the Chinese have their own peculiar varieties, all of them very different from those grown in this country. The favourite variety, " Peh-ts'ai," or Shantung cabbage, as

PEASANTS TRANSPLANTING RICE

foreigners have styled it, is more like a huge cos lettuce than a cabbage. This kind is grown everywhere, but attains its greatest perfection in the colder parts of China. In the Yangtsze Valley it is best when grown as a winter crop. Another striking variety is the white-ribbed cabbage, " Kin-ta-ts'ai," which is said to be peculiar to Szechuan. In addition to these some half-dozen other varieties are cultivated. Cabbages are eaten fresh or are preserved by salting and drying in the sun. From a European standpoint none is worth growing, being so very inferior in flavour to our own. The Roman Catholic priests have introduced the common European cabbage, but though its culture has spread widely the Chinese much prefer their own varieties. While the Chinese cabbages are all really referable to *Brassica campestris*, it is convenient to group them under *B. chinensis*. As a winter crop green kale, " Kan-kan-ts'ai," and dark-red kale, " Ts'ai-tai " are cultivated through the Yangtsze Valley. The young shoots of *Brassica juncea* and *B. campestris*, var. *oleifera*, are also used in the same way as kale.

The Chinese cultivate a great many gourds for food, the whole cucurbitaceous family being known under the general name of " Kua." Some are eaten raw, and others cooked. The male flowers, too, are eaten by the peasantry. The seeds of the water-melon are esteemed a great delicacy. They are slightly roasted, and are consumed in enormous quantities ; no banquet is complete without them, and over their gossip in tea-shops or restaurants, scholars and coolies alike regale themselves with these delicious morsels. Preserved in sugar, melon-seeds form a favourite sweetmeat. As a summer crop throughout the Yangtsze Valley the following cucurbitaceous plants are commonly cultivated : *Cucurbita Citrullus*, " Hsi-kua " ; *C. Pepo*, " Hsi-hu-lu " ; *C. moschata*, " Huo-kua " ; *C. maxima*, " Nan-kua " ; *C. ovifera*, " Sun-kua " ; *Cucumis Melo*, " Tien-kua " ; *C. sativus*, " Huang-kua " ; *Benincasa cerifera*, " Tung-kua " ; *Lagenaria vulgaris*, var. *clavata*, " Hu-tzu-kua " ; *L. leucantha*, var. *longis*, " Ts'ai-kua." When very young the fruit of *Momordica charantia*, " Ku-kua," is eaten, and when old is used as medicine. *Luffa cylindrica*, " Ssu-kua," is cooked and eaten when young ; when old the fibre is esteemed as medicine. *Lagenaria vulgaris*, " Hu-lu," is cultivated for its hard shells, which are converted into receptacles for holding water, oil, or wine. In addition to the above, several gourds are cultivated for their ornamental fruits, which are used for decorative purposes.

In the valleys and on the plains and low hills bordering them throughout the Yangtsze Valley and Yunnan, the Sweet Potato (*Ipomœa Batatas*) is the most important root crop. The crop is always cultivated on ridges and is grown from both old tubers and cuttings. Tubers are planted out in May, and cuttings from the shoots of these are inserted in July and early August, and produce a fine crop in October and November. The crop from the old roots is ready in August. Sweet potatoes are eaten after being boiled, baked, and dried in chips, and constitute a truly delicious dish. As they deteriorate by keeping, they are cut into slices, scalded, and then dried in the sun. The tubers are also macerated in cold water, and the resultant starch dried and made into vermicelli. In Hupeh the sweet potato is known as the " Hung-shao," in Szechuan as the " Pen-shao."

In the mountainous districts the sweet potato is displaced by the Irish potato, or " Yang-yü " (*Solanum tuberosum*), which, like maize, is another plant of American origin that has become a most important crop. It was introduced by the Roman Catholic priests at the time of a great famine some forty odd years ago. Its culture has spread enormously, and though it is despised by the rice-eating Chinese of the plains it has become a staple article of food with the highland peasantry. In the valleys it is cultivated as a late winter crop, in the mountains as a summer crop. Its culture is unfortunately but little understood ; it is always grown too thickly, and seldom if ever properly earthed up. Both red and white-skinned varieties are grown, but the flavour is usually very poor. The potatoes cultivated by the Buddhist priests on Mount Omei are justly celebrated, but the best I ever ate in China were grown by Sifan tribesfolk around Sungpan.

Two kinds of Yam are commonly cultivated, namely, *Dioscorea alata*, the " Chieh-pan-shao," which has enormously large, flat, branching tubers, and *D. Batatas*, " Pai-shao " ; both are cooked and eaten. Around Ichang the tubers of a third species are eaten. This species is known as *D. zingiberensis*, the "Huang-chiang," or Yellow Ginger. The tuber is bitter, and is valued chiefly as a medicine. Chinese yams do not equal the sweet potato in flavour, and are not so extensively grown. Around Chengtu *Pachyrhizus angulatus*, the "Ti-kua," is commonly grown. The white, firm-fleshed, turnip-like tubers are eaten either raw or cooked.

White turnips, "Lo-po," both the long and round kinds, are cultivated everywhere, but the flavour is very poor. Also the so-called red turnip, which really is a Radish (*Raphanus sativus*). All three are cooked and eaten when fresh, or preserved by being sliced and dried in the sun. *Brassica Napus*, var. *esculenta*, 'the "Ta-t'ou-ts'ai," is very generally cultivated, but I met with it most frequently on the Chengtu Plain. The whole plant is pickled and eaten with rice. The Szechuanese also cultivate most excellent Kohl-rabi (*Brassica oleracea*, var. *caulo-rapa*).

Two aroids, *Colocasia antiquorum* and its variety *Fontanesii*, " Kiang-tou," are very extensively cultivated for their tubers, which are cooked and eaten in various ways. Both are grown on ridges in flooded ground. The purple-coloured petioles of the "Kiang-tou" are sliced, pickled, and eaten. The flavour of the tubers of these plants is similar to that of the Jerusalem artichoke, but inferior. *Sagittaria sagittifolia*, " T'zu-ku," is cultivated in Szechuan and Yunnan, and the tubers are cooked and eaten in the same way as those of the Colocasia. The tubers of *Scirpus tuberosus*, " P'ei-chi," and the fruits of the Water Chestnut (*Trapa natans*), " Ling-chio," two very common aquatics, are esteemed valuable articles of food.

The Lotus-lily (*Nelumbium speciosum*), " Lien hwa," is cultivated both for its seed and its rhizome. These are used as food, but being expensive are luxuries enjoyed only by the wealthy. The fibres of the rhizome are used medicinally. Ginger (*Zingiber officinale*), " Seng-chiang," is very extensively grown. It is prepared for the table in various ways. From Canton, ginger preserved in sugar is exported in quantity to this country. *Amorphophallus konjac*, " Mo-yü," is sparingly cultivated throughout the Yangtsze Valley. The tubers are ground up with water and made into a curd-like compound. On Mount Omei and in north-west Szechuan this plant is more generally cultivated. The bulbs of *Lilium tigrinum*, the " Chia-peh-ho," are highly esteemed, and occur both cultivated and wild. The white bulbs of this lily are more expensive in China than they are in this country. When properly cooked these bulbs are not at all bad eating. They somewhat resemble the parsnip in flavour.

Of the onion family, Garlic, or " Ta-suan " (*Allium sativum*), and the common Onion; or " Ts'ong " (*A. Cepa*); are cultivated extensively. Garlic is highly esteemed. Onions are eaten as " spring onions," large bulbs being absolutely unknown. *A. fistulosum* is the Chinese Leek, " Chiu-ts'ai," and is very widely grown. The leaves are flattened and covered with earth to ensure blanching. The blanched leaves, " Chin-huang," are considered a delicacy. In the mountains *A. odorum*, *A. chinense*, and other species are common. These are culled and eaten by the peasantry. Szechuan, especially the more alluvial areas, produces remarkably fine Carrots (*Daucus Carota*), " Hung Lo-po." They are grown in large quantities and eaten with great relish. The Parsnip (*Peucedanum sativum*), " Uen-shui " are cultivated, but the roots are seldom thicker than a pencil. The whole plant is cooked and eaten.

Although in central and Western China quite a number of plants are grown for their oil, fully 75 per cent. of the oil commonly used is the product of two members of the cabbage family. After a careful investigation of the subject I have satisfied myself that the two plants in question are *Brassica juncea*, var. *oleifera*, and *B. campestris*, var. *oleifera*. The latter is the " Ta-yu-ts'ai " of the Chinese, the former the " Hsao-yu-ts'ai;" or " Ch'ing-yu." Both kinds are loosely designated " rape " by the foreigners resident in China ; but in my wanderings there I never met with the true rape plant. Throughout the entire Yangtsze Valley, during the winter months, enormous areas are given over to the cultivation of these two plants. Though the " Hsao-yu-ts'ai " is the earlier of the two, the other is the more extensively grown.

THE WAX-GOURD (BENINCASA CERIFERA)

These plants are in flower in February and March, and the crop is harvested in April. The seeds are crushed and steamed, and the oil obtained by expression. In Szechuan the use of the oil as an illuminant equals its culinary value. It also enters very largely into the composition of Chinese candles.

Oil is also expressed from the seeds of the Ground-nut (*Arachis hypogæa*), the Opium Poppy (*Papaver somniferum*), the Sunflower (*Helianthus annuus*), Cotton seed (*Gossypium herbaceum*), the Soy Bean (*Glycine hispida*), and members of the cabbage family, other than those already mentioned, notably the kales, and in the highlands from Flax seed, " Shan-chih-ma " (*Linum usitatissimum*). These oils are all used for cooking and lighting purposes and for adulterating the more valuable " Ts'ai-yu." With the exception, however, of the ground-nut, they are not extensively employed. In Hupeh and Szechuan, *Sesamum indicum* is cultivated sparingly as a summer crop. In Yunnan its cultivation is more general. The oil from its seeds is very highly esteemed, and commands a high price in the market. It is known as the " Hsiang-yu," or fragrant oil, and is eaten raw, mixed with cooked vegetables. From the seeds of *Perilla ocymoides* an oil, known as " Su-ma," and similar to sesamum oil, is expressed ; it is used in salads. This plant is, however, but very sparingly cultivated.

A large number of miscellaneous vegetables are used as food in various ways. Some are wild, but most are cultivated, and many of them are strange and novel to Europeans. A handsome if tasteless fruit, the Brinjal, " Chuei-tzu " (*Solanum Melongena*), is largely cultivated as a vegetable. The Chinese distinguish at least 5 varieties that differ from each other in colour, shape, and time of maturing. Some of them are truly enormous, often weighing 2½ lbs., and measuring 1 foot in length. They are in the market from June till October. The Tomato (*S. Lycopersicum*) has been introduced by foreigners, and in Yunnan is frequently met with semi-wild as an escape from cultivation. The Chinese, as far as my observations go, do not eat it themselves.

A small-fruited variety of the Chilli-pepper, " Ai-chiao " (*Capsicum frutescens*), is commonly cultivated, and is particularly happy in the dry, hot valleys of the Tung and Min

Rivers, where it is grown as an article of export for other parts of China. Both long and round (heart-shaped) forms of Capsicum (*C. annuum*) are cultivated in the plains, and especially the plain of Chengtu. These chillies and capsicums constitute the most important relish used by the Chinese. In a green state the latter are fried and eaten with rice and cabbage. When ripe they are pounded up in a mortar, and with water added form a sauce. Roasted and ground into meal they are used for seasoning purposes. The ripe chillies and capsicums are also boiled in oil, and impart to it their pungent flavour. Oil so treated will keep for an indefinite period. The true Chinese pepper, known as "Hua-chiao," is the ground-up fruit of *Zanthoxylum Bungei*. This is a thorny shrub cultivated everywhere in small quantities, but it is only in the Min Valley that I have noted it extensively grown for export.

As previously mentioned, bamboo shoots are eaten both fresh, dried, and salted. When cooked as a vegetable or made into a salad, these shoots are very fair eating, but it is ridiculous to compare them with asparagus, as some writers have done. In the warmer parts of China it is the young shoots of *Bambusa arundinacea* and *B. vulgaris* that are employed. They are also an article of export to other parts of China, and can usually be bought in a dried state in most of the large cities. In mountainous districts the young succulent shoots of other species of Bamboo are eaten. In the west, one of the commonest of these is the lovely *Arundinaria nitida*.

Celery (*Apium graveolens*), "Ch'ing-ts'ai," and Lettuce (*Lactuca Scariola*), "U'sen," are commonly cultivated. The celery is never bleached, and it is the stem of the lettuce rather than the leaves that is in request. The leaves and young shoots of the following plants are used as vegetables : *Cedrela sinensis*, "Ch'un-tuen shu"; *Pistacia chinensis*, "Huang-nien-ya"; *Chrysanthemum segetum*, "Tung-hao"; *Malva parviflora*, "Mao-tung-han-ts'ai"; *M. verticillata*, "Tung-han-ts'ai"; *Chenopodium album*, "Hui-t'ien-han"; *Acroglochin chenopodioides*, "Yeh-han-ts'ai"; *Ipomœa aquatica*, "Wêng-ts'ai"; *Anaphalis contorta*, "Tak'ing-ming-ts'ai"; *Coriandrum sativum*, "Yen-ts'ai"; *Taraxacum officinale*, "Ku-ts'ai"; *Beta vulgaris*, "T'ien-ts'ai"; *Lactuca denticulata*, "Wo-shengts'ai"; *Spinacia oleracea*, "Po-ts'ai"; *Crepis japonica*, "Huang-hua-ts'ai"; *Basella rubra*, "Juan-chiang-tzu"; *Celosia argentea*, "Chi-kung-hua"; and *Amaranthus paniculatus*, "Ya-ku."

The "Kao-sên" (*Zizania latifolia*) is very generally cultivated. Its succulent stem and very young inflorescence are cooked and eaten as a vegetable. From a European standpoint it is really very good eating. From the rhizome of the Bracken Fern (*Pteridium aquilinum*) an arrowroot-like substance called "Chüeh-fen" is prepared. In the mountains the young fronds of this fern are eaten by the peasantry. From the thick woody root of *Pueraria Thunbergiana* an "arrowroot" similar to the above is prepared. It is, however, in very little request, save in times of scarcity. The starchy roots of *Potentilla discolor* and *P. multifida* are also occasionally used for preparing a food-stuff.

The flowers of *Lilium Sargentiæ*, "Yeh-peh-ho," and of *Hemerocallis flava*, "Huang-hua-ts'ai," are eaten, as also are the yellow pea-like flowers of *Caragana chamlagu*. The mucilaginous seeds of *Plantago major*, "Ch'e-ch'ien-ts'ao," enter into the composition of a jelly, "Liang-fen," which is used in summer. The Chinese are very fond of several species of Fungi, and distinguish quite a number of edible kinds. Amongst their favourites are *Hirneola polytricha*, *Cantharellus cibarius*, *Tricholoma gambosa*, *Lactarius deliciosus*, and *Agaricus campestris*, the Common Mushroom. Seaweed is imported in quantity from Japan, and is on sale in the shops of all the larger towns and villages. From this Seaweed (*Porphyra vulgaris*) the Chinese prepare a very nutritious jelly.

The difficulty of tracing the original types of plants that have long been in cultivation and of affixing the correct scientific names to them is a very real one, and one that can be appreciated by all who have studied the history of our common garden plants. While in the foregoing pages I cannot hope to have altogether escaped error in this matter, I have used every means at my command to ensure accuracy.

CHAPTER VII

THE MORE IMPORTANT PLANT PRODUCTS

WILD AND CULTIVATED TREES OF ECONOMIC IMPORTANCE

CHINA is remarkably rich in raw economic products of vegetable origin, especially in oil, fat, and saponin-yielding fruits and seeds, lacquer-varnish, tannin, and dye-products, fibres and paper-making material. Some of these products are in increasing demand for export trade with the outside world, and will undoubtedly develop into great industries of the future. In this and the succeeding chapter is given an account of the more important of the products derived from central and Western China. This region is the source whence the majority of the raw articles are obtained that are exported from Hankow, the great trade entrepôt of the Yangtsze Valley.

One of the most important of all Chinese products is wood oil. This is obtained from the seeds of two species of Aleurites, a small genus of low-growing trees belonging to the Spurge family. The two species for the most part occupy distinct geographical areas, but both have been recorded as growing close together in the province of Fokien. In the south of China wood oil is the product of *A. montana*, which bears its flowers on the current season's shoots at the time when the leaves are expanded, and has an egg-shaped fruit, sharply pointed, and unevenly ridged on the outside. This is the "Mu-yu shu"—literally "Wood Oil tree" of the Chinese. In central and Western China it is *A. Fordii*, known as the "T'ung-yu shu"—literally "T'ung Oil tree," which produces this valuable oil. This latter species bears its flowers at the ends of the previous year's shoots before the leaves unfold, and has a flattened-round, apple-like fruit, only

64

slightly pointed, and perfectly smooth on the outside. These two trees have been very much confused by botanists, and it is well to emphasize their distinctive characters. The " T'ung-yu " is the more hardy tree of the two, and is much more widely distributed, furnishing fully nine-tenths of the so-called " wood oil " used in China and exported from thence. Within the last decade " wood oil " has attracted considerable attention in Europe and in the United States of America as a possible substitute for linseed oil, and it is annually imported into these countries in vastly increasing quantities. Chemists have investigated the products of these two trees, and find no appreciable difference in the oils.

The " Mu-yu " (*A. montana*) is common in the regions around Wuchou to the west of Canton, where it is chiefly used, and from whence it is exported to Hongkong and elsewhere. The trade is not large ; in 1910 it was estimated at 52,106 piculs.[1]

The " T'ung-yu " (*A. Fordii*) is abundant throughout the Yangtsze Valley from Ichang westwards to Chungking ; more especially it luxuriates in the region of the gorges and the contiguous hilly country up to 2500 feet altitude. It is essentially a hillside plant, thriving in the most rocky situations, and on the poorest of soils, where there is a minimum rainfall of 29 inches ; it will also withstand drought and a few degrees of frost. It is a quick-growing tree, seldom exceeding 25 feet in height and averaging less, with a much-branched, flat-topped head, 15 to 30 feet or more through, and is highly ornamental in flower and foliage. The flowers, produced in great profusion during April, are white, stained with pink and yellow markings, especially near the base. These are always followed by green, apple-like fruits, which ripen in September and are hidden amongst the large, glossy-green, heart-shaped leaves. Each fruit contains three to five seeds, which somewhat resemble shelled Brazil-nuts, but are much smaller.

The fruits break naturally in three parts when dead ripe, but they are invariably gathered before this period, and collected into heaps which are covered with straw or grass. Fermentation sets in and quickly disposes of the thin fleshy part of the

[1] One picul equals 133⅓ lbs.

fruit, after which the seeds are easily removed. The process
of extracting the oil is very simple. The seeds are first crushed
in a circular trough beneath a heavy stone wheel revolved by
horse or ox-power. The comminuted mass is then partially
roasted in shallow pans, after which it is placed in wooden
vats, fitted with wicker bottoms, and thoroughly steamed
over boiling water. Next, with the aid of an iron ring and
straw, it is made into circular cakes about 18 inches in diameter.
These cakes are arranged edgeways in a large press and, when
full, pressure is exerted by driving in one wedge after another,
thereby crushing out the brown, somewhat watery and heavy-
smelling oil, which falls into a vat below.   This " T'ung oil " is
packed in tubs and bamboo baskets, and is ready for export.
The yield is about 40 per cent. by weight of the kernels.   The
refuse cakes are used on the fields as fertilizers.

    " T'ung-yu " is the chief paint oil throughout the Chinese
Empire, being used for all outside woodwork ; as a " drier "
it excels linseed oil. The Chinese do not paint their boats,
they oil them, and the myriads of such craft which ply on the
Yangtsze and other rivers of China are all coated and the
upper works kept waterproof with this oil. The crude oil
boiled for an hour becomes a syrupy oil or " P'ei-yu," which
is used as a varnish for boats and furniture. Boiled for two
hours with the addition of certain mineral substances (" T'u-
tzu " and " T'o-shên "), a varnish called " Kuang-yu " is
produced which, when applied to silk gauze and pongees,
renders them waterproof.   " T'ung-yu " is also used as an
illuminant and as an ingredient in concrete ; mixed with lime
and bamboo shavings it is used for caulking boats. Besides
these, and dozens of other legitimate uses, " wood oil " is
also employed as an adulterant in lacquer-varnish. Lamp-
black produced by burning this oil or the fruit-husks is a
most important ingredient in the manufacture of Chinese
ink. The trade in " T'ung oil " is very large. From Hankow
in 1900 the quantity exported was 330,238 piculs, valued at
Tls. 2,559,344. In 1910 the trade had risen to 756,958 piculs,
valued at Tls. 6,449,421.

    I have given rather full details of this subject on account
of its great importance, and because its value is only beginning

THE TUNG-OIL TREE (ALEURITES FORDII) IN FLOWER: FRUIT

to be realized by the Western manufacturer. The U.S.A. Department of Agriculture has introduced *Aleurites Fordii* into its experimental stations, and expects to establish an industry in the production of " T'ung oil " somewhere in the United States of America. It is worthy of the serious attention of countries other than the United States of America. In South Africa, Australia, Algeria, Morocco, and other regions, for instance, this tree would probably thrive, and its experimental culture might with advantage be undertaken by the various Departments of Agriculture in those British Colonies and French Protectorates. Of all the varied economic vegetable products of China, the wood oils are pre-eminently of a kind to receive attention, with a view of establishing the industry in Colonial possessions.

Another member of the Spurge family yields the valuable Chinese " vegetable tallow " of commerce. This tree, *Sapium sebiferum*, occurs in all the warmer parts of China, and is remarkable for the beautiful autumnal tints of its foliage. This tree is known by several colloquial names—in southern China it is the " Chiu-tzu shu " ; in central parts the " Mou-tzu shu " ; in the west the " Ch'uan-tzu shu." It is a long-lived tree, growing 40 to 50 feet tall, and having a girth of 5 or 6 feet at maturity. In Hupeh, where the industry is well looked after, the larger branches are kept " headed in " to facilitate the gathering of the fruits. The fruits are three-celled, flattened-ovoid, about 15 mm. in diameter. When ripe they are blackish-brown and woody in appearance, and are either gathered from the trees by hand or knocked off by the aid of bamboo poles. After being collected, the fruits are spread in the sun, where they open, and each liberates three elliptical seeds, which are covered with a white substance. This covering is a fat or tallow, and is removed by steaming and rubbing through a bamboo sieve having meshes sufficiently small to retain the black seeds. The fat is collected and melted ; afterwards it is moulded into cakes, in which state it is known as the " Pi-yu " of commerce. After the fatty covering has been removed the seeds are crushed, and the powdered mass undergoes the same processes as are described for extracting wood oil. The oil expressed from the seeds is the " Ting-yu "

of commerce.   Very often no attempt is made to separate the
fat and the oil.   The seeds with their white fatty covering are
crushed and steamed together and submitted to pressure,
the mixed product so obtained being known as " Mou-yu."
The yield of fat and oil is about 30 per cent. by weight of the
seeds.   In China all three products are largely employed in
the manufacture of candles.   The pure " Pi-yu " has a higher
melting point than the " Ting-yu " or the mixture " Mou-yu."
All Chinese candles have an exterior coating of insect white
wax, but when made from " Pi-yu " only the thinnest possible
covering of wax is necessary (one-tenth of an ounce to a pound).
All three products of the Vegetable Tallow tree are exported
in quantity to Europe, where they are used in the manufacture
of soap, being essential constituents of certain particular forms
of this article.   Chinese vegetable tallow is an increasingly
important article of trade.   In 1910 some 178,204 piculs,
valued at Tls. 1,878,418, were exported from Hankow.

Every one is familiar with some form or other of the
lacquer-work of China and Japan, but the varnish employed for
lacquering has not yet found a market in Western countries,
owing to certain poisonous properties it possesses, and to the
want of knowledge as to the correct way of applying it.
Lacquer is prepared from a varnish obtained in its crude state
from *Rhus verniciflua*, the " Che shu " of the Chinese.   This
tree grows 25 to 60 feet tall, producing handsome pinnate
leaves, 1 to 2½ feet long, and large panicles of small greenish
flowers, which are followed by fruits rich in fatty oil.   It is
wild in the woods and abundantly cultivated along the margins
of fields throughout central China, especially in the mountainous areas of western Hupeh and eastern Szechuan, but
is much less common west of these regions.   Its altitudinal
range is from 3000 to 7500 feet, the optimum being 4000
to 5000 feet.   This tree, like the art of lacquering, was introduced from China into Japan in very early times, and is
commonly cultivated there to-day.   It is one of the many plants
which first reached Europe from Japan, of which country
it was erroneously considered native.

In China, Varnish trees are the property of the ground
landlord and not of the tenant who holds the land ; the varnish

is also claimed by the former. When the tree has attained a diameter of about 6 inches, tapping for varnish commences, and this operation is continued at intervals until the tree is 50 or 60 years of age. If the tapping is too severe, or the trees too young, injury or death ensues. The tapping operation is begun in late June or early July at a time corresponding with the opening of the flowers, and is continued throughout the summer. Oblique incisions from 4 to 12 inches in length, and about 1 inch in width, are made in the bark of the tree down to the wood, and the sap which exudes is collected in shells, bamboo tubes, and similar receptacles. Wooden pegs are driven into the trunk to facilitate climbing, in order to reach the main branches. The tapping is done early in the morning and the sap gathered from the receptacles into which it has flowed from the incisions each evening. In showery weather it dries rapidly, and often has to be scraped away. The sap continues to exude from the wound for about seven days, and then a fresh, thin slice of bark is removed, which causes another exudation. This is repeated seven times with an interval of about seven days between each operation, so that the work on each tree occupies about fifty days. After being tapped, the tree is allowed a period of from five to seven years to recover ; the old wounds are then reopened and fresh ones made. A arge tree yields from 5 to 7 lbs. of varnish. This, as it exudes, is pure white in appearance, but quickly oxidizes to greyish-white, changing to black. To prevent contact with the air the crude varnish is covered as soon as possible with layers of oil-paper.

Crude varnish furnishes only one colour, namely, black, and when applied to wood floors, or pillars, is the most indestructible varnish known. To obtain brown varnish " P'ei-yu " (*ante*, p. 66) is added, in the proportion of 25 to 50 per cent., to the crude varnish, according to the shade of brown required. The more " P'ei-yu " added, the quicker the varnish will dry. Red varnish is produced by adding cinnabar (mercuric sulphide) to brown varnish in about equal parts. Yellow varnish is obtained by adding to the " brown varnish " orpiment (arsenic sulphide) in slightly less than equal quantity.

Enormous quantities of raw varnish are exported from

central China to other parts of the Empire and to Japan. In 1910 the exports of varnish from Hankow totalled 15,424 piculs, valued at Tls. 1,043,434. This commercial product is frequently adulterated with wood oil. Three tests for adulteration are commonly employed—(1) Smell ; (2) the varnish is held up and allowed to drop, the strand of varnish will remain unbroken if it is pure, but will break if adulterated ; (3) placed on a sheet of soft Chinese paper, the varnish "runs," if it is adulterated, owing to the paper absorbing the oil adulterant. Everywhere in China this varnish is known to resident foreigners as "Ningpo varnish." The genesis of the name is interesting, since the substance itself is not produced in the neighbourhood of Ningpo, but is imported from Hankow and elsewhere. In the early days, when foreigners first settled at Shanghai, most of the carpenters employed to build houses for them were Ningpo men. For all indoor work—floors, pillars, and furniture—they employed this varnish, and foreigners promptly dubbed it "Ningpo varnish."

A peculiarity of "Ningpo varnish," or Chinese lacquer, to use its correct name, is that it hardens only in a moist atmosphere and remains in a tacky condition if exposed to sunlight and heat, the essentials in hardening copal varnish. In China it is applied only during cloudy weather when the atmosphere is surcharged with moisture or when a drizzle of rain is falling. For indoor work its drying is facilitated by hanging about the rooms cloths saturated with water. The kind used on ships contains "P'ei-yu" in almost equal parts, and this mixture dries rapidly even in moderately dry, hot weather. How important the knowledge of this peculiarity is may be gathered from the following fact. Many years ago an experimental consignment of "Ningpo varnish" was received in London. It was applied in the same way as ordinary copal varnish, in full sunlight and heat, with the result that it refused to harden, and remained "tacky," and the failure resulted in its being condemned as worthless !

The only change which takes place in the composition of the lacquer in drying at ordinary temperatures is the slow absorption of oxygen, finally amounting to 5·75 per cent. by weight of the original substance. Complete oxidation is found

THE LACQUER-VARNISH TREE (RHUS VERNICIFERA) 40 FT. TALL

to be due to the action of a ferment, to which the name *laccase* has been applied, which is only active in a certain humidity of the atmosphere. Quite recently, however, the presence of a special ferment has been questioned, and the absorption of oxygen attributed to an obscure chemical reaction depending on the presence of a compound of manganese with a proteid-like substance. Chinese lacquer, in a raw state, unfortunately possesses properties which are poisonous to many people, producing swellings and eruptions of the skin in the same way as does its close ally, the " Poison Ivy " (*Rhus Toxicodendron*). Certain people are immune, but this property will probably always militate against its use in Western lands. Perhaps the chemists will one day discover a means whereby this poisonous property can be neutralized or eliminated.

The fruit of *R. verniciflua* is shining, greyish-yellow, roundish and flattened on two sides, 6 to 10 mm. long. These when crushed and treated in a wedge-press in the same way as wood oil seeds, yield a fatty oil known as " Che-yu," which is used for making candles.

Trees belonging to three different families produce fruits rich in saponin which are in common use for laundry-work and other purposes. The most generally distributed of these " Soap trees " is *Gleditsia sinensis*, a handsome tree known colloquially as " Tsao-k'o shu," abundant throughout the Yang-tsze Valley up to 3500 feet altitude. It grows 60 to 100 feet tall, and has a thick trunk, smooth grey bark, a spreading head with massive branches, furnished with small, pinnate leaves and inconspicuous greenish flowers, unisexual or hermaphrodite in character. The latter are followed by pods or " beans " which, when ripe, are black, 6 to 14 inches long and $\frac{3}{4}$ to $1\frac{1}{2}$ inch wide. These pods are broken up and are in general use for ordinary laundry-work, producing a good lather in either cold or hot water. They are also used in the process of tanning hides. The saponaceous fat is contained in the pod itself, which is the only part utilized, the hard, flattened, brown seeds being discarded. It is probable that more than one species is included under the above name, for the Gleditsia family is in need of revision. In Yunnan another species which has much larger (20 inches) and wider pods is employed for the same purposes.

It is known as *G. Delavayi*. Around Peking a third species, designated *G. horrida*, occurs. A much rarer Soap tree, except in the vicinity of Kiukiang, is *Gymnocladus chinensis*, the " Yu-tsao-chio " of the Chinese, which is the Asiatic representative of the " Kentucky Coffee tree " of North America. This tree grows 50 or 60 feet tall, and though occasionally seen with a flat, fairly widespreading head, has usually only short branches ; the bark is smooth and light grey, the leaves much divided, often 2 feet across, pea-green in colour, and very handsome. The flowers, clustered, greyish without, purple within, are followed by flattened, brown pods, 3 or 4 inches long, and 1½ inch broad. These pods or " beans " are immersed for a time in hot water, which causes them to swell and become rounded in outline. Afterwards they are strung on short strips of bamboo and are then marketed. These swollen pods, colloquially " Fei-tsao-tou," are broken up and used in laundry-work, more especially for cleansing choice fabrics. They are also cut up into fine shreds and ground to a paste with sandalwood, cloves, putchuck, musk, camphor, etc., and thoroughly mixed with honey to form a perfumed soap called P'ing-she Fei-tsao (camphor-musk soap). This is a dark-coloured substance of the consistency of soft soap. It is used by women for cleansing their hair, and as a cosmetic for their hands and face ; also by barbers as a salve on the heads of their customers after shaving.

Yet another Soap tree is *Sapindus mukorossi*, colloquially known as the " Hou-erh-tsao." This occurs throughout the Yangtsze Valley up to 3000 feet altitude, growing 60 to 80 feet tall, with a huge trunk, smooth grey bark, and widespreading umbrageous head ; the pinnate leaves are 8 to 12 inches long. The flowers are small, greenish-white, produced in large terminal panicles, and are followed by shining brown, globose fruits about the size of a large marble. The fruits are used for washing white clothes, being considered for this purpose superior to the pods of Gleditsia. Each fruit contains a large, round, black seed. These are strung into rosaries and necklaces, which are much worn during hot weather.

During recent years the demand for vegetable products useful for tanning purposes has become unlimited. For certain purposes Chinese nut-galls furnish the finest tanning material in the world.   These " nut-galls " (" Wu-pei-tzu ") develop on the leaves of *Rhus javanica*, "Peh-fu-yang shu," as an excrescent growth due to protoplasmic irritation, occasioned by an insect (*Chermes*) which punctures the leaf to deposit its eggs.   The tree is of small size, and very abundant in the Yangtsze Valley up to 3500 feet, more especially in rocky places, producing panicles of white flowers in late August and September.   The galls are hollow and brittle, and vary considerably in shape and size, being more or less irregular, and 1 to 4 inches long.   In China they are used for dyeing blue silk and blue cotton cloth black.   The Occidental demand for nut-galls is greater than the supply, and the exports increase annually. In 1900, 24,800 piculs, valued at Tls. 454,584, were exported from Hankow ; in 1910 the exports from this port had increased to 53,784 piculs, valued at Tls. 936,234.

Another less common species is *R. Potaninii*, colloquially known as the " Ch'ing-fu-yang," which produces galls known as " Ch'i-pei-tzu."   These are used in Chinese medicine. The world is sadly in need of an indelible black ink, and chemists might well turn their attention to Chinese nut-galls in their quest for this treasure, since they possess possibilities worthy of investigation.

In the chapter on fruits reference is made to the cultivated Persimmon (*Diospyros kaki*), but it is necessary here to mention the feral form, known as " Yu-shih-tzu " (literally "Oil Persimmon ").   This wildling is abundant in the mountains of central and Western China up to 4000 feet altitude, where it forms a large tree 50 or 60 feet tall.   The fruit varies from flattened-round to ovoid, and from ¾ to 2½ inches in diameter.   It is always rich golden yellow in colour when ripe, and this colour best distinguishes the smallest fruited forms from its close ally *Diospyros Lotus*, " Kou-shih-tzu," which has flattened-round fruit, dark purplish coloured when dead ripe. To obtain the varnish oil for which this tree is esteemed, the fruit is plucked in July when about the size of a crab-apple, and still green.   By means of a wooden mallet the fruits are reduced to pulp, which is placed with cold water in large earthenware jars fitted with covers, and allowed to decompose.

The contents of these jars are stirred occasionally, and at the end of thirty days the residue of the pulp is removed and the resultant liquid, now a nearly colourless varnish, is poured into other jars. To give the varnish a warm brown tint, the leaves of *Ligustrum lucidum*, " La shu," sometimes erroneously called the " Tung-ching shu," are steeped in the jars for ten days or so, according to depth of tint desired. This varnish is used for waterproofing purposes generally, its principal use being in the manufacture of umbrellas. For this purpose it is applied as a gum varnish between the several layers of paper forming the screen of the umbrella, and serves to make them adherent as well as waterproof. When completed, the umbrella receives a thin outside coating of " Kuang-yu," or lustrous oil (*ante*, p. 66). Persimmon varnish is widely used, and is in great demand for the above purposes. It is produced in most parts of China, but scarcely figures as an article of export.

The art of making paper in China dates back to about the commencement of the Christian era. Previous to this, silk and cloth were employed for writing upon, but the early annals of the race were recorded on tablets of bamboo, and this latter method obtained in the days of Confucius (552–478 B.C.). What materials were first employed by the Chinese in paper-making are not known with certainty, but were probably bamboo or Paper-mulberry, " Kou shu " (*Broussonetia papyrifera*). A good case in favour of the latter could be made out, since the inner bark of this tree requires less preparation than bamboo culms. True paper money first originated in the province of Szechuan during the reign of the first emperor of the Sung Dynasty (A.D. 960). A certain Chang-yung introduced it to take the place of the iron money then in use, which was inconveniently heavy and troublesome. These notes were called " Chih-tsi " or " Evidences," and were apparently made from the inner bark of the Paper-mulberry. Marco Polo, speaking of Kublai Khan's mint at Peking, says, " He makes them take of the bark of a certain tree, in fact, of the Mulberry tree, the leaves of which are the food of the silkworms—these trees being so numerous that whole districts are full of them. What they take is a certain fine white bast or skin which lies

A SOAP-TREE (GLEDITSIA SINENSIS) 25 FT. TALL

between the wood of the tree and the thick outer bark, and this they make into something resembling sheets of paper, but black." The famous Venetian's error in calling this the silkworm Mulberry is pardonable enough, since the trees are very closely allied, and somewhat similar in appearance. Paper money is still made from the paper prepared from the bark of the " Kou shu," and the same paper, " P'i-chih " owing to its toughness, is used for wrapping up silver, for tags on silk goods, and as a lining between the fur or cotton and the outer fabric in fur-lined or wadded garments. The *B. papyrifera* occurs all over China up to 4000 feet altitude, and if left alone forms a much-branched tree 35 to 45 feet tall with a smooth dark grey bark. In a bush form it is abundant by the wayside and on cliffs. Most of the paper (which is called Kou-p'i-chih—literally bark paper) made from this tree and used in Western China comes from the province of Kweichou. In Hupeh the slender branches from young trees and bushes are cut into lengths, steamed in vats to facilitate the removal of the bark, which is converted into string and cordage.

The material from which the original India paper (a Chinese, not an Indian product), which came from Canton, was made is unknown. Possibly it was prepared from Ramie fibre (*Bœhmeria nivea*), but I venture the suggestion that it may have been obtained from the bark of *Broussonetia papyrifera*.

Bamboo supplies the material for the manufacture of all the better class papers used for printing and writing upon, papering windows, and a hundred and one other purposes. Several species are employed for this purpose, one of the commonest being *Phyllostachys heteroclada*. This bamboo is abundant in central and Western China, especially in alluvial areas near streams up to 4000 feet altitude. It grows 12 to 18 feet tall, with fairly slender dark green culms ; commonly it forms extensive groves. The stems are cut into lengths, made into bundles, and immersed in concrete pits, being weighted down and kept under water by heavy stones. After three months they are removed, opened up, and thoroughly washed. Next they are restacked in layers, each layer being well sprinkled with lime and water, holding potash salts in solution. After two months they are well retted. The fibrous mass is then washed to remove the lime, steamed for fifteen days, when it is removed, thoroughly washed, and again placed in concrete tanks. The mass is next reduced to a fine pulp with wooden rakes, and is then ready for conversion into paper. A quantity of the pulp is put into troughs with cold water and mucilage prepared from the roots of *Hibiscus Abelmoschus*. An oblong bamboo frame, the size of the desired sheet of paper, having a fine mesh, is held at the two ends by a workman and drawn down endways and diagonally into the liquid contents, which are kept constantly stirred in the trough. It is then gently raised to the surface, and the film which has collected on the top is deposited as a sheet of moist paper when the frame is turned over. After the surplus water has drained away from the mass of moist sheets of paper the whole is submitted to pressure. It is then dried either in kilns or in the sun, according to quality, the sun-dried being the inferior. Since much water is necessary in the process of paper-making the mills are always erected alongside streams.

The more common paper in daily use is made from rice-straw by a similar but less intricate and quicker process. The stems of a reed (*Imperata arundinacea*, var. *Kœnigii*), known as "Mao-ts'ao," and common in many parts of Western China, are also used locally in the manufacture of paper, being frequently mixed with rice-straw.

Chinese "rice-paper," so called by foreigners, is prepared from the pith of *Tetrapanax papyrifera*, a shrub closely allied to the common Ivy of Europe, and colloquially known as "T'ung-ts'ao." This plant has handsome palmate leaves, and stems filled with a pure white pith. This pith is cut, using a rolling, circular motion, by means of a sharp, heavy knife, into thin sheets. Formerly much of this cutting was done in Chungking, the raw material being imported from the province of Kweichou. Rice-paper is used by Chinese artists for painting upon, and also in the manufacture of artificial flowers.

Sericulture and silk-weaving are among the most important industries of Szechuan. Nearly every part of the province produces silk, but there are certain well-defined areas in which the industry is famous—for example, Kiating Fu, Chengtu Fu, and Paoning Fu. Hosie [1] estimates the annual production of raw silk at lbs. 5,439,500, valued at Tls. 15,025,230. This industry has been exhaustively dealt with by Hosie (loc. cit.) and others, and I propose here only to mention briefly certain trees, the leaves of which the silkworms are fed upon. The overwhelming proportion of Szechuan silk is produced by the "worm" of *Bombyx mori*, the common domesticated species, which is fed principally on the leaves of the White Mulberry (*Morus alba*), known as the "Sang shu." The Mulberry tree is abundantly cultivated up to 3000 feet altitude, and in the more populous parts of the province a traveller is seldom out of sight of groves of this tree. The trees are kept low by pollarding to admit of the leaves being easily gathered, but little attention is otherwise given them. Since the suppression of opium cultivation the officials have turned their attention to improving and extending the sericulture industry. The finest Chinese silk is produced in the neighbourhood of Hanchou in the Chekiang province, where a broad-leaved and particularly fine Mulberry is cultivated (*M. alba*, var. *latifolia*), for the purpose of feeding the silkworms. The recently established Bureau of Agriculture at Chengtu Fu, and magistrates in charge of certain districts, have introduced the Hanchou Mulberry in the hope of improving the local product. During the last two or three years there has been a considerable increase in the area devoted to sericulture, and there is a possible danger of over-production. More attention might well be paid to the spinning of the yarn in order to produce a more even thread, which would result in a smoother and finer woven fabric.

Around Kiating Fu the infant silkworms are fed for the first 22 days of their lives on the finely chopped leaves of *Cudrania tricuspidata*, the "Tsa" or "Cha shu," a low-growing tree (very often only a bush), closely allied to the Mulberry, with thorny branches and dark green, tough leaves. For the succeeding and final 26 days they are fed on the Mulberry. By feeding first on the Cudrania leaves, it is claimed that the

[1] *Report on Province of Ssuch'uan*, p. 61.

worms produce more silk of a tougher and more durable quality. Hosie [1] was the first to discover and make known this interesting fact to the outside world, and subsequent observers have confirmed his statements.

Around Paoning Fu in the north, and Kikiang Hsien in the south, a certain amount of silk is obtained from the worm of *Antherœa pernyi*. This species feeds on the leaves of various scrub oaks, and being bivoltine, produces two crops a year. Several species of Oak are concerned, including *Quercus variabilis*, *Q. serrata*, *Q. Fabri*, and *Q. aliena*, all of which, though they attain to the dimensions of trees, are commonly met with from 2000 to 4000 feet in the form of bushes covering the hillsides. This Oak-feeding silkworm was introduced from the province of Shantung many years ago, and the industry is much more important in Kweichou province than it is in Szechuan. This " Wild-silk," as it is called, differs from ordinary silk in its harder texture and is spun from dry cocoons, whereas ordinary silk is spun from cocoons lying immersed in boiling water.

In 1907, near the hamlet of Lu-yang-ho, alt. 2500 feet, in the north-west corner of Fang Hsien, I chanced upon several plantations of *Ailanthus Vilmoriniana*, grown for feeding the worm of *Attacus cynthia*. The trees were all young saplings. This was the only place in my travels where I saw this particular kind of sericulture practised. In parts of north-eastern China I understand it is more general, the species there employed being the ordinary *Ailanthus glandulosa*, the " Ch'ou-ch'un shu " of the Chinese, and " Tree of Heaven " of foreigners.

[1] *Three Years in Western China*, p. 21.

CHAPTER VIII

THE MORE IMPORTANT PLANT PRODUCTS

CULTIVATED SHRUBS AND HERBS OF ECONOMIC VALUE

CHINESE agriculture is mainly devoted to the production of food-stuffs for local consumption, the surplus being disposed of by sale and the proceeds invested in the necessities or luxuries of life which cannot be produced locally. Nevertheless, in the more fertile parts of the Empire, certain economic crops other than those for culinary purposes are grown expressly for sale or exchange. This is particularly true of the rich province of Szechuan, where a number of such products are produced, as will be seen from the brief account of the more important which follows.

Had this been written five years previously it would have been necessary to give considerable space to the Opium Poppy, but so vigorously has the edict for the suppression of this crop been promulgated that only a brief notice of it is necessary. When the Imperial Decree, prohibiting the cultivation and consumption of opium throughout the Chinese Empire within a period of ten years, was published on 20th September 1906, I confess to having been one of those who considered it a fatuous effort calculated to accomplish nothing, though well-meaning enough. It seemed impossible that such a gigantic task could be accomplished in such a brief period of time. Public sentiment was obviously in favour of the Decree, but to certain provinces, for example Szechuan and Yunnan, the export of opium represented their principal source of income. That Indian opium could be dispensed with and none be inconvenienced save the wealthy opium-smoking connoisseurs living in the prosperous coast ports, was perfectly clear to any one who had travelled in Western China. In 1908 the area under poppy in Szechuan was far greater than it had ever been before. In 1910 I traversed this province from east to west and north to south, and was amazed to find the whole industry of poppy-growing blotted out of existence. Except in a few out-of-the-way places, where it was grown by stealth, the cultivation had ceased. What has happened since the end of 1910 I do not know, but from what I saw brought to pass in a couple of seasons, together with the undoubted general disfavour in which opium-smoking was viewed by the people, I am constrained to think that the poppy and opium will disappear from China as it has done from Japan. The problem before officials, and more especially those of the western provinces, is to find a source of revenue to take the place of that formerly derived from opium. In 1904 Hosie estimated the production of opium in Szechuan at 250,000 piculs. In 1910 some 28,530 piculs of opium (produced in Szechuan, Yunnan, and Kweichou provinces), valued at about Tls. 29,000,000, passed through the port of Ichang. In 1909, 51,817 piculs passed through this port. Formerly the exports of opium alone from Szechuan nearly sufficed to cover the imports of cotton-yarn and piece-goods, commodities essential to the people of that province.

The literature on Chinese opium and opium-smoking in China is enormous, and with exception of what is written above; I desire to add only three significant facts, which, if known, are not generally appreciated. For the benefit of those who believe, and those who do not believe, that India, abetted by the British Government in times past, is responsible for the opium vice in China, I would mention that (a) opium has been known in China since the Tang Dynasty (A.D. 618), and was cultivated in Szechuan for medicinal purposes during the closing years of that Dynasty (circa A.D. 900); (b) the pipe used for smoking opium in China is of a design peculiar to the country itself; (c) the races of Poppy cultivated in Western China are allied to the races grown in Persia and quite distinct from those grown in India.

It is known that in early times the peach, orange, and silk travelled from China by the ancient trade route across Central Asia to Persia, from whence they reached Europe. Is it not, therefore, reasonable enough to suppose that the opium poppy

FIELDS OF THE OPIUM POPPY

may have travelled from Persia to China by this same overland route ?

The poppy is (or was) a winter crop in Szechuan, being garnered in April and May in ample time to prepare for the rice-crop. No other crop even remotely approximating the pecuniary value of opium can take its place.

Several plants yielding fibres valued for textile and cordage purposes are grown in China. In Szechuan the most important of these is the true Hemp (*Cannabis sativa*), colloquially known as " Hou-ma." This crop is abundantly cultivated around Wênchiang Hsien and P'i Hsien. It is a spring crop, the seeds being sown in February and the plants harvested the end of May and beginning of June, just as they commence to flower. The stems are allowed to grow thickly together and reach 8 feet in height. The culms are reaped, stripped of their leaves, and often the fibre is removed there and then. More commonly, however, the stems are placed in pits filled with water and allowed to ret for a few days ; they are then removed, sun-dried, stacked in hollow cones, surrounded by mats, and bleached by burning sulphur beneath the heaps. After these processes the fibrous bark is stripped off by hand. The woody stems that remain after the bark has been removed are burned, and the ashes resulting, mixed with gunpowder, enter into the manufacture of fire-crackers. Hemp, or " Hou-ma," is the best of the fibres produced in Western China for rope-making and cordage purposes generally. It is also used locally for making grain-sacks and coarse wearing apparel for the poorer classes. Quantities are used in the city of Paoning Fu for these latter purposes. It is in great demand on native river-craft and is largely exported down river to other parts of China. It is this hemp that is principally exported from Szechuan. True Hemp (*Cannabis*) is an annual and is grown as a summer crop in the mountains for the sake of its oil-containing seeds. Hemp oil is expressed and used as an illuminant and is said not to congeal in the coldest weather. In Hupeh it is known as " T'ang-ma."

Another annual plant cultivated for its fibre is *Abutilon Avicennæ*, the T'ung or T'uen-ma of Szechuan and Hupeh. This plant is widely cultivated as a summer crop in Western China up to 3000 feet altitude. The fibre is of inferior quality and is used locally for making cordage and in caulking boats, but is less valuable than that of the true Hemp (*Cannabis*) and less important as an article of export from Szechuan. Jute (*Corchorus capsularis*), colloquially known in Szechuan as " Huang-ma;" is very sparingly cultivated on the Chengtu Plain and elsewhere. It is not exported from the province.

The brown fibre from the leaf-bases of a palm, *Trachycarpus excelsus*, known in Hupeh as " Chung-ma," is the " coir-fibre " of the Yangtsze Valley. This " coir " is made into bales and exported down river from Szechuan in quantity. It is used for rope-making, mats and mattresses, brushes, is converted into rude raincapes, and is an all-round useful fibre.

The most important textile plant in China is the much-discussed China-grass, Ramie, or Rhea (*Bœhmeria nivea*). This member of the nettle family is both wild and cultivated in all the warmer parts of the Middle Kingdom up to 4000 feet altitude. It is a herbaceous perennial and grows 3 to 6 feet tall ; the leaves, broadly ovate, abruptly cuneate, or truncated at base, have dentate margins and are silvery on the underside. In Hupeh the wild plant is called " Ch'u-ma," the cultivated plant " Hsien-ma." In Szechuan the cultivated plant is also known as " Hsien-ma " and occasionally as " Yuang-ma." These various colloquial names are most perplexing and are almost hopelessly confused.

In Szechuan small patches of this " China-grass " are to be found around nearly every peasant's home. South-west of Chungking and also north of Lu Chou in several districts, it is cultivated on a very extensive scale. Much of the fibre is woven into " grass-cloth " and used locally. A certain amount is also exported down river. Szechuan " grass-cloth " is rather coarse and very much inferior to that produced in parts of southern China. It is not a prominent export from the west. In 1910 the exports from Hankow amounted to 120,034 piculs, valued at Tls. 183,332. This is classified in the Customs returns as Ramie fibre, and does not include that woven into grass-cloth.

RAMIE OR CHINA-GRASS (BOEHMERIA NIVEA)

Cotton-cultivation is a comparatively recent industry in China, having been introduced early in the eleventh century A.D., from Khoten. It met with strong opposition from those interested in the production of silk, China-grass, and other fibres, and was not fairly established until some time during the Yuan (Mongol) Dynasty (A.D. 1206–1386), when a public-spirited woman, Lady Hwang, distributed seeds throughout Kiangan, now the great cotton region of China. Chinese cotton has a notoriously short staple, but is strong and durable. It has undoubtedly become exhausted from lack of any attempt at seed-selection and from long cultivation in the same regions. Cotton-cultivation should receive early attention from the new Government, and seeds of standard varieties from India, Egypt, America, and elsewhere might be secured and experimentally grown. There is no question but that China could produce cotton infinitely superior to the present product if new and suitable varieties were obtained and properly cultivated.

Very little cotton, " Mien-hwa," is grown in Western China; and cotton-yarn and cloths are the great import into Szechuan. The value of foreign imports into Chungking is about Tls. 20,000,000, five-sixths of which is made up of cotton manufactures, the bulk of which comes from India.

Before the importation of mineral-oil from foreign countries became general, the only lamps in use were vessels filled with vegetable-oil and fitted with rush-wicks. These "rushlights" are still in common use in the west, more especially among the poorer classes. The wick consists of the pith of *Juncus effusus*, known as " Teng-ts'ao," which is widely cultivated for this purpose. The plant grows 3 to 6 feet tall and is also largely employed in the manufacture of matting and mats used under bed-mattresses and on divans. It is expressly cultivated for this purpose in parts of Szechuan, the principal seat of the matting industry being Sui Fu, where both whole and split rushes are used. In Yunnan *Scirpus lacustris*, " Pu-chih-ts'ao," which produces stems 6 to 8 feet tall, cylindrical at base, gradually tapering upwards and becoming obtusely triangular near the summit, is used for mat-making. It is also sparingly employed for the same purpose in Szechuan, where, however, it is chiefly used by the shop-keepers as string.

Rice-straw is largely used for making bed-mattresses and sandals and to a less extent for ropes. Wheat-straw is braided and used for making large, wide-brimmed summer hats. Certain districts like Shuangliu Hsien, near Chengtu, are famed for their straw-braid, but the industry is of local importance only.

Tobacco (*Nicotiana Tabacum*), called " Yen," was probably introduced into China from America, contemporaneously with maize—just when is a matter of dispute, but some sinologues consider it was about A.D. 1530. It is cultivated all over China, and nowhere within the Middle Kingdom are finer tobacco leaves produced than in Szechuan. Within the rice-belt tobacco is a spring crop, the seeds being sown in late October and the crop harvested by mid-June. In the maize-belt it is grown as a summer crop but not extensively. The districts of Chint'ang and P'i Hsien, on the Chengtu Plain, are noted for their tobacco. In these districts one crop only is taken from the plants, but in the warmer parts of the province contiguous to the Yangtsze River, three crops are secured before the plants are ploughed under.

Tobacco leaves are prepared in three ways : (1) the large leaves are dried on screens, kept flat, packed into bales to form " Ta-yen," or large tobacco ; (2) the smaller leaves are dried in the same way to form " Erh-yen," or second tobacco, which, when treated with Chinese rape-oil and red-earth (Tu-hung) is pressed and shaved into fine shreds and used for smoking in water-pipes; being known as " Shui-yen," or water tobacco ; (3) " So-yen," or cord tobacco, prepared by cutting off the leaves with a piece of the stem to form a hook, by means of which the leaves are suspended under the eaves of houses or from rafters indoors and allowed to dry, naturally shrivelling and curling in the process. This " So-yen " is rolled into rough cigars, which are inserted into the bowl of long-stemmed pipes and smoked. It is also exported from Szechuan. In the mountains up to 9000 feet altitude the small-leaved *Nicotiana rustica*, " Lan-hwa-yen," is sparingly cultivated for local use.

This receives no preparation beyond being dried in the sun, and naturally the quality is very inferior.

Undoubtedly the climate and soil of Szechuan are suitable for the growth of tobacco, but, unfortunately, the Chinese methods of curing the leaf are slovenly in the extreme, with the result that the prepared article is of low-grade quality. The Chinese are, unfortunately, fast becoming a nation of inveterate cigarette-smokers. Much of the local Szechuan tobacco could be used in the manufacture of cigarettes were proper factories erected. This has been done at Hankow and elsewhere, where cigarettes are manufactured from tobacco grown in the neighbourhood and in near-by provinces.

Sugar is a very important crop in Western China, and enormous quantities are produced in certain parts of Szechuan, where it is cultivated in the drier regions of the rice-belt up to 2500 feet altitude. Two kinds of Sugar-cane (*Saccharum officinarum*) are grown : (1) red-cane, used for chewing ; (2) white-cane, for the extraction of sugar. The Red-cane (*S. officinarum*, var. *rubricaule*) produces culms 8 feet tall, an inch or more in diameter, and is treated as an annual. The canes are cut as they mature and sold as required ; the canes that remain at the end of the season are taken up by the roots in November, cleaned and stored in earth-burrows until required for sale. About the end of March portions of these canes are laid lengthwise under the soil, and young growths that develop from each joint in due season constitute sugar-canes. These culms are dark red-purple outside, yellowish within, very firm, and rich in sugar.

The White-cane (*S. officinarum*, var. *sinense*) is treated as a perennial, producing two or three crops before being renewed. It grows 10 to 15 feet tall, with " long-jointed " stems nearly an inch in diameter. This is much more extensively grown than the " red " variety, and supplies nearly all the sugar used locally or exported from the province. Chinese methods of crushing the cane are very imperfect, and their refining processes are most primitive. The canes contain a high percentage of saccharine, and the industry, if perfected, could become of vast importance.

Sugar has been cultivated in China from time immemorial.

It is everywhere called " T'ang," and generally supposed to commemorate the T'ang Dynasty (A.D. 684–907), one of the most famous in Chinese history. Sugar, however, was known to the Chinese at least as early as the second century B.C., and is mentioned in a poem which was written sometime between A.D. 78 and 139.

Formerly the Chinese used only vegetable dyes for their silk and other fabrics, and it is much to be regretted that in China, as elsewhere in the world, these are being rapidly displaced by aniline dyes derived from coal-tar. The latter are more convenient to handle, but unfortunately the colours are not " fast." The coal-tar product is on sale in every town and market village in Western China, made up in small bottles and imported from Germany.

The only dye-plant at all extensively grown in Szechuan to-day is *Strobilanthes flaccidifolius*, " T'ien-hwa," which produces an " indigo." In certain parts of the Chengtu Plain this is grown in quantity, and the same is true of the district of Mien Chou and elsewhere, but its cultivation is on the decline. It is planted on ridges which are kept flooded between. When the plants are about 3 feet tall they are cut down and the leafy shoots placed in concrete pits full of cold water. After steeping for about five days the stems are removed, leaving a green-coloured water. Slaked lime is placed in the water to precipitate the indigo. The water is allowed to drain off, and the dye is found deposited at the bottom of the pit.

Around Shasi, in Hupeh, *Polygonum tinctorium* is cultivated as the source of an " indigo " which is there used for dyeing cotton cloth.

As a red dye Safflower (*Carthamus tinctorius*), " Hung-hwa," was formerly very extensively grown, but it is only occasionally met with to-day, though still esteemed for dyeing the more costly silk fabrics. The flowers of the Balsam (*Impatiens Balsamina*), colloquially " Chih-chia-ts'ao," are similarly used and valued.

Yellow dyes are obtained from turmeric, the root of *Curcuma longa*, still extensively grown in Chienwei Hsien on the Lower Min River, and from the flowers of the Huai shu (*Sophora*

A FIELD OF TOBACCO

japonica), a common and widely dispersed tree. Another, but much more rare tree (*Kœlreuteria apiculata*), is known by the same colloquial name, and the flowers are used for the same purpose as those of the Sophora. The fruit of *Gardenia florida*, " Chih-tzu-hwa," is used for dyeing certain woods yellow, and also as a yellow colour in paint.

Green dyes were formerly obtained from the leaves of *Rhamnus davuricus*, known as the " Tung-lu," a very common Chinese species of Buckthorn, extremely variable in the size and shape of its leaves and abundant as a thorny bush by the wayside everywhere up to 4000 feet altitude. Another species (*Rhamnus tinctorius*), " Chiao-lu-tsze," was also employed for the same purpose. These have been almost totally displaced by aniline dyes.

As mentioned on page 73, the gall-nuts (Wu-pei-tzu) produced on the leaves of *Rhus javanica* are extensively employed for dyeing fabrics—more especially silk—black. With this dye it is essential that the material be first dyed blue. The burr-like cupules of two very common species of Oak (*Quercus serrata*, *Q. variabilis*), known as the " Hwa-li " and " Hwa-k'o-li " respectively, are also commonly employed as black dye for silk-yarn and fabrics. In this case it is immaterial what the original colour may be. The curious cone-like fruits of *Platycarya strobilacea*, colloquially known as the " Huan-hsiang shu," are in general use as a black dye for cotton-yarn and cotton goods generally. Pine soot, obtained by burning the branches of the Common Pine (*Pinus Massoniana*), is also employed as a black dye for cotton goods.

As a dark brown dye and tanning agent the tubers of a yam are commonly used in Yunnan and are exported in quantity to Tonking and elsewhere. It is probably *Dioscorea rhipogonioides*, a species common in Formosa, where it is called " Shu-lang " and much used for dyeing and tanning fish-nets. In western Hupeh the root-bark of *Rosa Banksiæ*, called " Hu-p'i," is used for this purpose.

Both sesamum and soy bean are cultivated extensively in Western China, but for local consumption only. The large exports of these products that pass through Hankow are brought down by the Peking-Hankow railway. Szechuan is capable of growing enormous quantities of these valuable plants, but cheaper and better facilities for transport are necessary before the products can become articles of external trade. When the much-discussed Hankow-Szechuan railway is *fait accompli* the raw products of the west will be available as articles of export, and a much-needed stimulus given to the agricultural industries of the regions concerned.

CHAPTER IX

TEA AND "TEA-YIELDING" PLANTS

THE TEA INDUSTRY FOR THE THIBETAN MARKETS

THE most widely known product of China is, of course,
Tea, "Ch'a," which to-day is very extensively cultivated
in India, Ceylon, and Java, and also experimentally in
several other countries. In China the value of this plant has
been appreciated from very early times. It is known to have
been cultivated in Szechuan during the early Han Dynasty
(202 B.C.–A.D. 25). However, it was not in general use among
all classes before the sixth and seventh centuries A.D. Very
early in the seventeenth century tea first became known in
Europe, having been brought from China by Dutch traders.

The Tea plant (*Thea sinensis*) is considered to be a native
of Assam, whence it was long ago introduced and cultivated in
China. Augustine Henry, in 1896, received through a Chinese
collector whom he had trained specimens of undoubted *Wild
Tea*. Henry writes:[1] "Hitherto the Tea plant has been found
wild only in Assam, the cases of its spontaneity recorded from
China being very doubtful. In all my trips in Szechuan and
Hupeh I never met with it. The present specimens are above
suspicion, coming from virginal forest (in the extreme south-
south-east corner of Yunnan) and at an immense distance from
any tea-cultivation, the nearest being P'uêrh, 200 miles west.
Bretschneider, in his *Botanicon Sinicum*, Part II, p. 130, has
some remarks on the antiquity of tea in China. It is probable
that it was found wild in these southern provinces which did
not form a part of the ancient Chinese Empire, and I dare say
it will be found wild in these mountains from Mengtse to
Szemao. *It is not probable at all that tea came from so far away*

---

[1] *Kew Bulletin*, 1897, p. 100.

as Assam." I have italicized Henry's concluding statement, with which I most emphatically agree. As recorded in Vol. I, Chapter VIII, I discovered specimens of the Tea plant in north-central Szechuan growing in situations which left no good reason for regarding them as other than spontaneous. However, in view of the long-cultivated character of this shrub I prefer to regard them as " probably wild plants." It is worthy of note that growing in the same locality I found wild plants of the Tea Rose (*Rosa indica*) in some quantity. The Tea plant is an evergreen, belonging to the rain-forest area of the temperate zone in China. This represents the rice-belt throughout the Yangtsze Valley, which has long since been cleared in all but the most precipitous places to make way for cultivation. This fact would account for the present absence of the Tea plant in a wild state in these regions.

The great tea-growing districts for export trade with the Occident and for consumption within China itself are in the middle-eastern parts of the Empire. The export trade in this commodity has declined enormously during the last quarter of a century. Some 60 years ago the tea industry was introduced as a business into India and Ceylon, with the result that to-day these countries supply the greater portion of the world's demand. Antiquated methods of cultivation and preparation, absence of co-operation amongst the growers, and heavy taxation, are responsible for the decline of the Chinese product. It is true that Chinese tea is in quality and delicacy of flavour far ahead of Indian and Ceylon teas, but tea-drinkers generally have acquired a taste for the rougher, dark-coloured teas, and China's conservative methods are killing what was once her greatest export industry. Hankow is to-day the great tea-mart of China, the trade being largely in the hands of Russians. Large factories have been established expressly for the purpose of preparing teas for the Russian market, Indian and Ceylon teas being imported for blending purposes. In 1910 the exports of tea from Hankow were valued at Tls. 18,423,474.

With the ordinary tea industry of eastern China we are not further concerned, but in the west a specialized form of this obtains which merits a detailed description. Tea is grown all over Szechuan for provincial consumption, but in the western parts it assumes much greater magnitude, being there grown and specially prepared for the Thibetan market. The one great export from China to Thibet is tea, either in the form of compressed " bricks " or " bales." The subsidy given by the Chinese Government to the Thibetan authorities at Lhassa and elsewhere in Thibet is also paid in tea.

To the Thibetans tea is an absolute necessity of life, and deprived of this astringent they suffer in various ways. That astringency is one of the properties most desired is evidenced by the fact that the bark of Oak trees is ofttimes used when tea cannot be obtained. The ordinary everyday meal of these people consists of tea mixed with a little butter and salt. To this mixture roasted barley-meal is added, and the whole is kneaded to the consistency of dough, in which condition it is eaten. Buttered tea is also their national beverage. To the European palate this concoction as prepared by the Thibetans bears only the remotest possible resemblance to " tea." I have tried it often but never succeeded in persuading myself to like it.

Much has been written on the possibility of Indian tea-planters having a share in this tea trade with Thibet. From the close proximity of Assam to Lhassa and south-eastern Thibet generally, one would suppose that the difficulties would not be very great, yet the trade has made little progress. The opposition of the Lamas and the obstinate conservatism of the people are very real difficulties in the way. There is also another and equally important factor which should not be lost sight of, namely, the nature and quality of the tea that is in demand. Now it is safe to say that the veriest sweepings from the Indian tea factories would make better tea than that partaken of by the average Thibetan; but this is not the important point. To secure a share of this trade Indian planters must be prepared to supply the Thibetans with the kind of article to which they are accustomed, and not with something different, even though it be of a superior quality. The trade is very considerable and worth striving after; further-more, there is no reason why it should not be increased. I was travelling on the Chino-Thibetan frontier during the time of the British Expedition to Lhassa, and discussed with Chinese merchants interested in the Thibetan tea trade the possibility of India taking a share in the trade. It was very evident that they greatly feared Indian competition, and were keenly alive to the possibilities of it. From Darjeeling to Lhassa is only about 30 stages (350 miles approx.), whilst from Tachienlu the journey occupies over three months. The physical difficulties of the route are greater on the Chinese than on the Indian side, yet the people of Lhassa still draw their tea-supply from China. And further, Chinese tea, apart from that taken in exchange for musk, skins, wool, gold, and medicines, was, until very recently, paid for by the Thibetans in Indian rupees.

The brick-tea prepared for Thibet is a totally different article from that prepared in Hankow for the Russian market. It is also so totally different from ordinary Chinese tea that some have supposed it to be the product of a distinct plant. My wanderings in Western China led me through the tea-producing areas and the markets which supply the commodity to the Thibetans, my observations, therefore, may be of interest and value.

The two great trade-marts for China and Thibet are Tachienlu, in the west of Szechuan, and Sungpan, in the extreme north-west corner of that province. The official route to Lhassa passes through Tachienlu, and this town is the mart for southern and central Thibet, including Lhassa, Chamdo, and Derge. The mart for the Amdo and Kokonor regions generally is Sungpan. At this latter town the trade is purely one of barter, tea being taken in exchange for furs, wool, musk, and medicines. The tea for the two markets is prepared very differently, grown in distinct localities, and is best discussed separately.

The tea for the Tachienlu market is practically all grown within the prefecture of Yachou Fu, more especially in the mountainous districts to the north-west and south of the town. The manufacturing business is controlled by the Government and provincial authorities, who issue a definite number of licences to establishments in the towns of Yachou, Mingshan, Yungching, and T'ienchüan—all within the Yachou prefecture. The independent department of Kiung Chou, a little to the north-

BRICK TEA FOR THE TACHIENLU MARKET

east of Yachou, also has a share in this trade, but there the licences are all issued by the Imperial Government and are not connected with the provincial authorities at Chengtu. The industry is a very ancient one, the plant itself having been grown in this vicinity since the dawn of the Christian era.

To supply the licensed establishments the peasants and farmers cultivate the tea plant. The culture extends up to 4000 feet altitude, the bushes being planted round the sides of the terraced fields on the mountain-sides. Very little attention is given them and they are usually allowed to grow smothered in coarse weeds to a height of from 3 to 6 feet. Less frequently are the bushes kept free of weeds. During the summer months the leaves and young twigs are plucked off and placed, handfuls at a time, in heated pans for a few minutes, and then spread out in the sun to dry. They are then collected into large sacks or into loose bales and carried down to the towns and villages, where they are purchased by agents of the tea establishments. Occasionally the bushes, when they have become old, are cut down, the branches dried in the sun, and afterwards tied into bundles and carried down for sale. The very young leaves and tips of the shoots are commonly gathered by the growers and prepared into tea for home consumption and local trade, the old coarse leaves and branches being considered good enough for the Thibetans.

I visited a brick-tea factory in Yachou, where I observed the following processes of manufacture : The sacks of leaves and bundles of leafy sticks, after they had fermented for a few days, were taken in hand by women and children who picked off the leaves and shoots, sorting them into four grades, each grade being determined by the size and age of the leaves. The sticks, often 1 to 2 inches in circumference, after the leaves have been removed, were chopped small by means of a large knife fixed in a block of wood. Mixed with coarse leaves and sweepings these chopped-up sticks constitute the fourth grade. A small packet of the very worst of this grade is inserted in the ends of each bamboo-cylinder as a gratuity to the repackers and muleteers at Tachienlu.

A certain British consul has likened this brick-tea to " crows' nests pressed into cakes." This aptly describes the product so far as the fourth quality is concerned, but the first quality, prepared at Yachou, is really very good tea. I was surprised at the care and attention bestowed on its manufacture, the processes being as follows : After the leaves had been sorted and graded they were steamed in a cloth suspended over a boiler. The steamed mass was then put into collapsible moulds, together with a little of the dust from smashed sticks and leaves which had been treated with glutinous rice-water to make it cohere, and then the whole was submitted to great pressure. When the mould was removed the tea was in the form of bricks (Chuan), each measuring 11 inches by 4 inches and weighing 6 English pounds. After being dried for three days, these bricks are wrapped in paper on which the maker's trade-mark is stamped, a patch of gold-leaf of minute proportions or a plain piece of red paper to denote the quality being also enclosed. Four of the bricks are then placed end to end in a plaited bamboo-cylinder, and after this has been fastened at the ends the tea is ready for transit. These bamboo-cylinders, when filled with tea, are called " Pao " ; they weigh 25 lbs. and measure about 4 feet in length. They are carried on the backs of coolies to the town of Tachienlu, where they pass into the hands of Thibetans. The bricks of the finer quality teas and those intended for the interior of Thibet and distant Lhassa, are removed from the bamboo-cylinders and repacked, 12 together, in raw yak-hides, with the hair inside and the free edges neatly sewn together. The inferior quality teas are largely consumed in eastern Thibet and are not repacked. From Tachienlu the packages are carried on the backs of yak and mules to their destination.

The " Pao " packed in Yachou city always weigh 18 catties [1] (24 lbs. English), but in other places they vary according to quality, being either 12, 13, 14, 15, or 16 catties, each town having its own particular weight for the different qualities. Tea from Yachou city and Yungching Hsien follow the main road ; that from Mingshan and T'iench'uan a by-road. Both routes converge at the town of Luting chiao, and there pay toll on crossing the river. Either route is terribly difficult, and one marvels how such loads can be carried by men over

[1] Catty = 1⅓ English pounds.

such fearfully mountainous roads. The average load consists of 10 pao of 18 catties each. But loads of 12 and 13 pao are very common, and on several occasions I have seen men carrying 20 pao. These, however, only weighed 14 catties each, but even then the total weight of the load was 370 English pounds !

The distance between Yachou and Tachienlu is about 140 miles (probably less), and the journey for coolies laden with tea occupies 20 days. Although the work is so inhuman, thousands of men and boys are engaged in this traffic. With their huge loads they are forced to rest every hundred yards or so, and as it would be impossible for the carrier to raise his burden if it were once deposited on the ground he carries a short crutch, with which he supports it when resting, without releasing himself from the slings.

For each pao carried from Yachou to Tachienlu the carrier receives 400 cash (about a shilling in English money). Out of this he has to keep himself and pay for his lodgings. Nevertheless, the pay is really good for the country, and it is this extra remuneration that tempts so many to engage in this work.

It is very difficult to obtain accurate information as to the extent of this transfrontier tea-trade, but statistics culled from various, more or less reliable, sources, show that at the lowest estimate some 5400 tons of brick tea worth approximately £150,000 enter Tachienlu annually.

Tea for the Sungpan market is grown in two distinct localities, in the west and north-north-west of the Chengtu Plain respectively. Each district has its own peculiar mode of packing the product. In the west it is grown in the mountains bordering the banks of the Min River in the district of Kuan Hsien. A centre of the industry is the market village of Shui-mo-kou, some 90 li beyond the city of Kuan Hsien itself. This tea is not pressed into bricks after the manner of that for the Tachienlu market, but is made into rectangular bales some $2\frac{1}{2}$ feet by $2\frac{1}{2}$ feet by 1 foot each, weighing 120 catties (160 English pounds), and covered with bamboo-matting. A considerable quantity of this tea finds a market among the Chiarung tribes, the distributing centres being Monking Ting and Lifan Ting. The mountainous regions of An Hsien and Shihch'uan Hsien constitute the north-western tea district, the principal centre being the market village of Lei-ku-ping within the district of An Hsien. The prepared product, however, all passes through Shihch'uan Hsien, and is controlled by specially appointed officials. The tea prepared in this region is packed in oval bales, each weighing 65 to 70 catties (about 90 English pounds), encased in the usual bamboo-matting.

The routes by which Kuan Hsien and Shihch'uan Hsien teas travel converge at Mao Chou, an important town situated on the left bank of the Upper Min River, six days' journey south of Sungpan Ting. To Mao Chou the tea is mostly carried by men, two small or one large bale being the usual load. From Mao Chou to Sungpan mules and ponies are largely employed for transporting it, their loads being twice the weight of those carried by men. Both women and men, however, are also engaged in the carriage of tea from Mao Chou northward, and the merchants constantly complain of insufficient means of transport.

The preparatory processes undergone by the tea destined for the Sungpan market are less intricate than those described for brick tea. The leaves and young branches are gathered, panned, and dried in the sun. The panning process is some-times omitted, and very commonly the bushes and their overgrowth of coarse weeds are cut together, dried in the sun, and tied into bundles. The leaves are collected into sacks or bales, and with the bundles of leafy sticks carried down to the market villages and sold to tea establishments. The manufacturers allow the leaves to ferment in heaps for a few days, and afterwards submit them to a rough sorting. The sticks are chopped up with the coarse leaves and steamed over a large pan of boiling water. The moist, heated mass is then firmly pressed into bales, covered with matting, and allowed to dry.

The tea is practically all of one quality, and very little superior to the most inferior kind entering Tachienlu. Cheap-ness is the main consideration, a bale of 120 catties being valued in Sungpan at Tls. 8. This trade is a monopoly in

A TEA PLANTATION

the hands of five establishments, which pay to the provincial government at Chengtu a fixed tax of about 1 cent. per catty. Payment is done by purchasing permits called " Yin piao," which bear the official stamp. Each permit covers a bale of 120 catties or two smaller ones, and costs Tls. 1·20.

Whereas at Tachienlu the tea passes directly into the hands of Thibetans, at Sungpan it remains in the hands of the five tea establishments. These are owned by Mohammedan Chinese, who, in addition to carrying on a considerable local trade, have trusted agents travelling all over north-eastern Thibet bartering tea for furs, wool, musk, medicines, and other Thibetan commodities.

The tea-trade of Sungpan is an improving one, but it is practically impossible to obtain reliable figures of its volume. There are, of course, Chinese Official Returns stating the number of " Yin piao " sold annually, but where official peculation is so general such returns are notoriously untrustworthy. Piecing together information gathered during my three visits to Sungpan, I suggest that the tea-trade averages about £75,000 annually.

From all sources the total annual value of the tea exported from China to Thibet is about a quarter of a million sterling. On paper this may not appear very great, but if the sparse population of Thibet and the difficult means of intercommunication be duly considered, it will be seen that the trade is really a very considerable one. Indian teas cannot compete with the Chinese product in central and northern Thibet, but around Lhassa and in southern Thibet generally, they ought to command a market.

In all the larger medicine shops in Szechuan and, incidentally, elsewhere in the Empire, a product known as " P'uêrh tea " is on sale. It is packed in circular cakes, flat at top and bottom, about 8 inches across, and covered with bamboo leaves fastened by strips of palm leaves. This tea is grown in the Shan states, largely in the district of I'bang, and is the product of a variety of the true Tea plant (*Thea sinensis*, var. *assamica*). It takes its name from P'uêrh Fu, a prefecture in southern Yunnan, and the trade entrepôt of that region. The leaves, after the necessary preliminary processes, are steamed and pressed into the cakes, in which form they are easily transported. P'uêrh tea has a bitter flavour, and is famous as a medicine all over China, being esteemed as a digestive and nervous stimulant. It also finds its way into the wealthy lamaseries of Thibet, where its medicinal properties are highly appreciated.

Although a beverage known as tea is partaken of throughout the length and breadth of the Middle Kingdom, it is by no means all infused from the leaves of the genuine tea plant. In the mountainous parts of central and Western China many substitutes are employed by the peasants, who seldom taste the real article. In western Hupeh the leaves of several kinds of Wild Pear and Apple, grouped under the colloquial name of " T'ang-li-tzu," are used as a source of tea and exported to Shasi for the same purpose. The infusion prepared from these leaves is of a rich brown colour, very palatable and thirst-quenching. It is called Hung-ch'a (red tea), and is in general use among the poorer classes in the west.

The leaves of *Pyracantha crenulata*, the Chinese " Buisson ardent," are also in common use as a source of tea. This ever-green is everywhere abundant up to 4500 feet altitude, and is known as the " Ch'a kuo-tzu," literally, " Tea shrub." Like its European relative it produces a wealth of scarlet fruit in autumn. The leaves of several species of Spiræa (*S. Henryi*, *S. Blumei*, *S. chinensis*, and *S. hirsuta*) are less commonly used as tea, being known as " Tsui-lan ch'a." The leaves of the Weeping Willow (*Salix babylonica*) are occasionally employed as tea, and in the Upper Min Valley chips of willow-wood are likewise used. I have drunk all these various " teas," but that infused from these willow-chips was the worst, being decidedly weak and nasty !

In the chapter on Mount Omei mention is made of the sweet tea prepared from the leaves of *Viburnum theiferum*. The leaves of the common White Mulberry steamed, mixed with cabbage-oil, and pressed into cakes, constitute " Ku-ting-ch'a " (bitter tea). The infusion prepared from this is drunk in hot weather and esteemed as a cooling beverage.

The product known as tea-oil is not produced by the tea plant, but is expressed from the seeds of *Thea Sasanqua*, known as the "Ch'a-yu kuo-tzu," a relative of the true tea plant, from which it may be readily distinguished by its hairy shoots. It is a shrub, common as a wild plant in the sandstone ravines of north-central Szechuan. In parts of eastern China it is abundantly cultivated for the sake of its oil, but in the west I only met with plantations in the district of An Hsien. It is, however, reported as being cultivated in the department of Kiung Chou and elsewhere. The oil is used to adulterate cabbage-oil, and by Chinese ladies as a dressing for their hair. The refuse cake is valued as a fertilizer, and when applied to rice fields is said to destroy the earth-worms which often attack the young rice plants.

CHAPTER X

INSECT WHITE-WAX

NEXT to sericulture the most important industry in the prefecture of Kiating is that concerned with the production of insect white-wax or " Peh-la." This product has attracted the attention of many travellers, and has been often discussed before. It possesses several peculiarly interesting features, and cannot be omitted from any account of the economic products of western Szechuan. It is produced by a scale-insect (*Coccus pela*), and is deposited on the branches of an Ash (*Fraxinus chinensis*) and a Privet (*Ligustrum lucidum*); the scale-insects are bred in one district and transported to another for the production of the wax. All this sounds very simple, yet it has taken nearly five centuries to establish these facts. According to Chinese historians insect white-wax first became known to the Chinese about the middle of the thirteenth century. Nicolas Trigault, a Jesuit missionary, wrote some account of the industry in parts of Eastern China in the year 1615. During succeeding centuries several accounts of it were published, but it was not until 1853, when Mr. William Lockhart, of Shanghai, sent specimens of crude wax to England, that the wax-producing insect became scientifically known in England. In the crude wax a number of dried, full-grown bodies of the female insect were discovered, and were identified by Westwood as a new species of Coccus. Robert Fortune, in his travels around Ningpo in 1853, had noted the industry, and stated that " the tree on which the wax is deposited is undoubtedly a species of Ash." In 1872 the illustrious Baron Richthofen wrote of the production of insect white-wax in Western China, a fact not previously known to the people of the Occident.

In 1879 Mr. E. C. Baber made a lengthy report on the

white-wax industry of Western China from observations near
Fulin. Unfortunately, this talented observer possessed no
botanical knowledge, and, being misled by vernacular names,
he increased if anything the mystery which shrouded the
botanical aspect of the subject.

In 1884 Mr. (now Sir Alexander) Hosie, then Consular
Agent at Chungking, undertook, at the instigation of the Kew
authorities, the thorough investigation of the subject. He
travelled through the principal wax-producing districts of
Szechuan, collected specimens of the two host-plants and of
the wax itself, noted the mode of culture and the preparation
of the commercial white-wax. The two host-plants were
identified by the Kew authorities as *Ligustrum lucidum* and
Fraxinus chinensis, the first named being the tree on which
the insects breed and the latter the tree on which the wax is
deposited. There can be little doubt that the Ligustrum is
the natural host of the wax-insect, and much of the difficulty
in elucidating the subject was due to the fact that this tree
has two or three different vernacular names. In central and
Western China it is usually designated the " La shu " (Wax
tree) or " Ch'ung shu " (Insect tree), but it is occasionally,
and particularly in the eastern provinces, called the " Tung-
ching shu." This last name simply means " Winter-green
tree," and is usually applied to *Xylosma racemosum*, var.
pubescens, a tree commonly planted around shrines and
tombs. Many wild guesses were made as to the identity
of this " Tung-ching shu," and with each guess the subject
became further involved.

The districts of Omei Hsien and Hungya Hsien, both within
the prefecture of Kiating, are the headquarters of the wax-
producing industry, but the insects are bred in the Chiench'ang
Valley, in the prefecture of Ningyuan Fu, nearly 200 miles
distant. A few insects are bred near the town of Chienwei
Hsien, a day's journey to the south of Kiating, but these are
said not to produce so much wax or of such good quality as
those from the Chiench'ang Valley.

The insects develop during the winter months, and the
cone-like scale or " gall " is ready for removal about the end of
April, being then full of the minute eggs of the insect. So far

as my observations go, they indicate that it is always on the
Privet that the insect breeds, but Baber asserts that either
tree will serve, and this is probably true.

Several of these cone-like scales, full of eggs, are wrapped
together in thin paper bags, which are arranged in airy crates
and carried by porters with all possible speed to the city of
Hungya, where they are disposed of to the farmers. During
the month of May hundreds of coolies are engaged in this traffic.
The larvæ hatch out quickly, more especially if the season is
hot and early, in which case the travelling is mostly done at
night by the aid of lanterns. The journey of nearly 200 miles
over difficult mountain roads is accomplished in six days.
Aided by relays, the porters who carry these insects cover
30 to 40 miles per day ; in ordinary circumstances 20 miles
a day is a high average for porters in the west.

For the production of the wax it is immaterial whether the
Ligustrum or Fraxinus is used. Some districts favour the
latter, others the former; very frequently the two trees are
grown side by side. The trees are planted round the edges of
the fields, and are polled some 5 or 6 feet from the ground. The
lateral shoots, which develop from the polled heads, are always
one or more years old ere the insects are placed on them. The
propagation of these trees is effected by taking thick branches,
slicing off a portion of the bark and a little of the wood, and
surrounding the incised area with a ball of mud and straw.
Roots form in the ball of mud, and the branch is then
severed from the parent tree, and is planted at the side of a
field, where it quickly develops into a tree.

In the wax-producing area of the Kiating prefecture
myriads of these pollarded trees are cultivated by the farmers
and peasants. Previous to the arrival of the insects in May,
the branches on which it is intended to place insects are
denuded of their laterals along the basal half of their length.
The cultivator, having purchased his insects, wraps loosely a
few cones in a broad leaf and suspends these tiny " bags "
among the branches of either Fraxinus or Ligustrum trees, or
of both. The larvæ quickly hatch out and crawl up into the
tree and ascend to the leaves, where they remain for fourteen
days until " their mouths and limbs are strong." During this

INSECT WHITE-WAX AND TREE (LIGUSTRUM LUCIDUM)

period they are said to " moult," casting off " a hairy garment which forms in the earliest larval stage." After this period the insects descend to the naked branches, on the underside of which they attach themselves and commence at once to deposit wax. During this early stage heavy rains and wind are much dreaded, since they dislodge the insects, and consequently ruin the business for the season. The deposit of wax, which at first looks very like hoar-frost on the branches, continues up to the latter end of August. (The Chinese reckon 100 days from the time of suspending the insects in the trees.) The deposit is always heaviest on the underside of the branch, and seldom extends equally all round it.

About the end of August the white coating is scraped from the branches (very often the branches are cut off) and thrown into boiling water. The wax is dissolved and floats on the surface of the water. It is collected by being skimmed off, and whilst in a plastic state is moulded into thick saucer-shaped cakes. The insects sink to the bottom of the vessel containing the boiling water, and are collected and thoroughly crushed to express every particle of wax before being finally flung to the pigs.

The wax excretion has been attributed to disease, but in the light of present knowledge it seems feasible to regard it merely as a device on the part of Nature to protect the insect from its enemies. The Chinese idea is that the insects live on dew, and the wax perspires from their bodies !

The natural enemy of the wax-insect is a species of " Lady-bird," which breeds with them and preys on the larvæ. The Chinese designate this enemy " Wax-dog" (La-gho). After the larvæ have hatched out, the farmer visits his trees in the heat of the day, and belabours their stumps with a club for the purpose of dislodging this foe.

The co-operation which obtains in this industry between two separate and distinct districts has led to much confusion. The explanation seems to be that owing to peculiar climatic conditions the insect breeds freely in Chiench'ang Valley, and for similar reasons deposits wax freely in the Kiating pre-fecture. At any rate, it is obvious that one cannot have wax and insects too, since to obtain the former it is necessary to kill the latter by immersion in boiling water. I am convinced that the co-operation or mutual dependency is simply one of self-interest on the part of both districts.

Insect white-wax bears a close resemblance to spermaceti, but is much harder. It is colourless and inodorous, or nearly so, tasteless, brittle, and readily pulverisable at 60° F. It is slightly soluble in alcohol, and dissolves with great facility in naphtha, out of which fluid it may be crystallized. It melts at about 180° F., floats in water, and is said to harden by long immersion in cold water.

The wax is largely used in the manufacture of Chinese candles, a little being mixed with the fats and oils employed in their manufacture ; a thin coating is also applied to the outside of the candles. The best candles contain 2½ ounces to the lb., inferior ones not more than 1 ounce. Since the ordinary fats and oils melt at about 100° F., the advantage of an outer coating of white-wax with its high melting-point is obvious. In paper-shops insect white-wax is largely employed to impart a gloss to the higher grades of paper. In medicine-shops it is universally used as a coating for pills, and is itself supposed to possess medicinal properties. It is also employed as a polish on jade and soap-stone ware and on the more delicate articles of furniture, to give lustre to cloth, and is made into ornaments of Buddha ; but its primary uses are in the manufacture of candles and in paper-glazing.

The annual output varies considerably, the industry being almost entirely dependent upon suitable climatic conditions. In poor seasons 50,000 piculs is an average crop, whereas in very favourable years it is more than double this quantity. Formerly the prefecture of Paoning produced a fair amount of white-wax, but the industry has there become neglected of recent years. To-day practically the whole supply of Western China is produced in the Kiating prefecture.

In spite of the increased consumption of foreign candles and kerosene oil, the demand for insect white-wax remains steady, and the industry concerned with its production shows very little sign of decline. In Western China, owing chiefly to difficulties and dangers of navigation on the Yangtsze, and the consequent heavy freights, foreign goods are an expensive luxury enjoyed only by the wealthy. With the advent of railways vast changes will certainly take place, and this interesting insect-wax industry may at some future date become extinct.

CHAPTER XI

SPORT IN WESTERN CHINA

A TRAVELLER in Western China who is fond of the sport will, in season and from time to time, have opportunity of enjoying some good rough shooting. During my travels in that land I have had with dog and gun some splendid days—days which afford keen pleasure to look back upon. My aspirations in the matter of shooting never extended beyond the Pheasants, though at odd times I have shot a few River Deer, Muntjac, and, of course, Hares. But my extensive and prolonged journeys in the more mountainous parts of China have afforded me great opportunities of gaining a knowledge of the game-fauna of Western China.

During the years 1907–09, the expedition under my charge paid particular attention to the fauna, and amassed a collection of some 3135 birds, skins of 370 mammals, and specimens of various reptiles and fishes. My associate on this particular expedition, Mr. Walter R. Zappey, had especial charge of the collecting work in this department, and it speaks volumes for his enthusiasm, untiring energy, and skill that in so short a time he succeeded in making such a magnificent collection. The specimens he obtained are, through the munificence of John E. Thayer, Esq., Lancaster, Mass., U.S.A., preserved in the Museum of Comparative Zoölogy at Harvard College. The entire collection has been worked up by various specialists, and the results published in the *Memoirs* (vol. xl. No. 4, August 1912) of that Museum.

This expedition gave me facilities for acquiring an intimate acquaintance with the fauna of Western China, and enables me to submit to readers much first-hand information relative to this subject.

Mr. Oliver G. Ready in his *Life and Sport in China*, and Mr. H. T. Wade in his *With Boat and Gun in the Yangtsze Valley*, with other writers, have given accounts of the game-fauna found in the more accessible parts of eastern China, but I am unacquainted with any work giving a general, descriptive account of the game animals and birds of the mountainous parts of central and Western China. In *The Middle Kingdom*, by S. Wells Williams, a brief notice of the fauna of China is given, but this was written long ago, and is based largely upon Chinese evidence and hearsay, and in consequence cannot be regarded as either complete or accurate. Unfortunately in the new edition of this work, published in 1900, very little revision of the chapter dealing with the fauna was attempted, and not much new matter was added.

Since the subject under discussion is a large and comprehensive one, it is simplest, perhaps, to divide it into two distinct parts, one dealing with the birds and the other with the mammals.

BIRDS

There is a great variety of game birds and wild-fowl found all over China ; moreover, this land is the headquarters of the Pheasant family. This latter fact is alone sufficient to make China of particular interest to all sportsmen, even had she no other attraction to offer. A Chinese Pheasant, commonly called the " Chinese Ring-neck " (*Phasianus torquatus* in the widest sense), was, as the world of sport well knows, long ago introduced into Europe and crossed with the native bird. To-day practically all the Pheasants bred in England for purposes of sport have more or less of this Chinese blood in them. The Mongolian Pheasant (*P. mongolicus*), a hardier bird, with lighter plumage in the female and young birds, is now being commonly bred for shooting purposes. In the United States of America the same obtains, but to a much more limited extent. Other Chinese Pheasants such as the " Reeves," " Golden," " Amherst," " Tragopan," etc., have been introduced, but have not been so readily acclimatized, and are comparatively seldom seen outside the aviary.

The country of western Hupeh, of which the city of Ichang may be regarded as the " Gate," constitutes a natural boundary for members of the animal and vegetable kingdoms. The flora and fauna found east of this region are, generally speaking, totally different to that found to the westward. The explanation is to be found in the character of the country. At Ichang commence the series of mountain ranges which, rising higher westward, finally culminate in the mighty snowclad ranges of the Chino-Thibetan borderland. Enclosed within this mountain-system is the Red Basin of Szechuan which is described in Vol. I, Chapter VI. This highly cultivated Basin again constitutes a barrier, and very few of the game birds or animals are common to both its eastern and western boundaries. East of Ichang for 1000 miles to the coast lay the vast alluvial plains and flats of the Yangtsze Valley. Here and there mountain ranges crop out like islands in the ocean, and so long have these elevations been isolated that they support in the main a peculiar flora and fauna. The surprisingly restricted range of the component species is one of the most interesting facts in Chinese natural history.

Of true Pheasants (*Phasianus*) some six species occur in the region with which this work is concerned. Each of these species occupies its own particular geographical area. But it must be admitted that the modern tendency of systematic ornithologists to split up species into subspecies and varieties based on very slight variations renders the subject complicated and difficult. Indeed to such an extent is subdivision carried that one is sometimes inclined to think the actual differences exist on paper only.

The low foot-hills, which commence some 30 miles east of Ichang, constitute the western limit of the common Ring-neck Pheasant of middle eastern China and the Yangtsze Valley in particular (*P. torquatus*, var. *kiangsuensis*). It is likewise the eastern boundary of a Pheasant in which the ring is usually quite absent (*P. holdereri*). The " Ring-neck " is essentially a bird of the plains, whilst the other is a mountain bird, adapted to more austere conditions of life.

The bird of the plains is almost semi-aquatic in its habits, breeding in swamps and places more generally associated with waterfowl than pheasants. In the reed-bed region bordering the Tungting Lake I have, on several occasions, enjoyed good shooting, and it was always in the wet marshy places that the birds were most plentiful. Winter Snipe are common in this same region, and a right and left Pheasant and Snipe is commonly obtainable. On my first shooting trip in the reed-bed region I was ignorant of the aquatic habits of this Pheasant, and a friend and I worked all the dry, likely-looking places for the best part of a day with most discouraging results, until we accidentally plunged into a swampy region and found the birds. Water, food, and cover are everywhere the three essentials for pheasants of all kinds, but this bird of the plains seems to have a stronger predilection for the first-named than does any other species. The probable explanation is that in the marshes it enjoys greater protection from enemies, both two and four footed, than in the dry, open, and highly cultivated plains.

In times past some extraordinary bags of this Pheasant have been made, and records of such are given in Mr. Wade's book. The bird is still quite common in the Yangtsze Valley, but natives, shooting for the market supply of the various treaty-ports, and especially that of Shanghai, have cleared many of the best districts known to foreigners. Phenomenal bags are no longer obtainable, and each year the foreign sportsman has to go farther afield if he wishes to enjoy good shooting.

The cunning of the " Ring-neck " is proverbial, and the bird is so well known that description is unnecessary. The white ring round the neck and the white eyebrows are constant features that distinguish this bird from other species. The average measurement of the male is 32 to 34 inches tip to tip.

The common Pheasant of the mountains of central China (*P. holdereri*) is about the same size as the " Ring-neck." The head and neck black with bluish and green reflections, occasionally with a more or less complete white neck-ring ; breast purplish, abdomen black ; sides dull yellow, each feather having a black spot near the tip ; upper back dull yellowish, feathers notched and margined narrowly with

black; lower back bluish-slate, tail broadly barred, sides of
upper tail-coverts light chestnut; length about 32 to 34 inches
tip to tip. The more broadly-barred tail-feathers, absence
of white eyebrows, and the usual absence of the white neck-
ring distinguish this bird from the common " Ring-neck."

If the Pheasant of the plains is notoriously cunning, his
confrère of the mountains is equally so, and the nature of his
haunts aids him considerably in escaping his enemies. A
common habit when hunted is for him to work his way quickly
to the top of a steep hill and take wing on the crest. A study
of his habits is necessary before much success attends one's
efforts at shooting this Pheasant. The haunt of this mountain
bird is the woods, copses, and scrub-clad mountain-sides, but
he is seldom found in quantity other than in close proximity to
cultivation. Just how far west this bird ranges I do not
know, but he has not yet been authentically recorded west of
the eastern limits of the Red Basin. His headquarters is
undoubtedly western Hupeh and southern Shensi. His
altitudinal range in this region is only limited by cultivation.
In the mountains a favourite food of this bird is the fruit of
many Rosaceous shrubs, particularly that of Cotoneaster.
Scrub Oak retaining its warm brown foliage through the winter
is general throughout this region, and is in winter a favourite
haunt of this bird. In heavy snow he seeks the forest,
especially that composed of evergreen trees.

This Pheasant is strong on the wing and capable of carrying
away a lot of shot. Much of the shooting in mountainous
country is snap-shooting, and one's powder should " hit hard "
or the bird is round the corner out of sight, and probably lost.
Shooting this mountain bird is much finer sport than that
afforded by the plains species. Every bird secured in the
mountains is earned, and this combined with the bracing air
gives additional zest and pleasure to the sport. I have spent
some very pleasant days after this bird, and though I have
never made a big bag I have enjoyed some enviable sport.

Around Ichang, where this Pheasant occurs in sparse
and ever-decreasing numbers, the hills are covered with a
" Spear grass " (*Heteropogon contortus*), called " Hung-tsao "
by the Chinese. The seeds of this annoying grass are barbed,

and they drill their way through clothing deep into the flesh, from whence they are not readily extracted. Their power of penetration is truly marvellous. Ordinary cloth, such as serge, flannel, and khaki, is useless against them—they will even penetrate through the leather tongue of a shooting boot ! Stout duck and drill, starched and glazed, are the only kinds of woven material that will resist them, and then only as long as the material remains dry. In this spear-grass country only smooth-haired dogs are useful and capable of facing the cover. In the more mountainous country, three or four days removed from the river, the most useful kind of dog to cover the country is probably the Spaniel.

In the mountains of south-western Szechuan west of the Min River Valley and as far north as lat. 31° N. the common Pheasant is Anderson's (*Phasianus elegans*). This lovely bird differs from Holderer's Pheasant in having a dark green instead of purple breast and terra-cotta instead of dull yellow sides ; the rump more pronounced slaty-grey with green reflections ; the tail is shorter, and the bird, which averages 29 to 30 inches tip to tip, is smaller in all its parts. The average weight of full-grown cocks is 2½ lbs. An imaginary line connecting Kuan Hsien and Tachienlu roughly marks its northern range ; southwards it extends through western Yunnan to the borders of Burmah. Its altitudinal range is from 2500 feet up to 10,000 feet, or even higher where cultivation obtains. Around Tachienlu, at alt. 8000 to 9500 feet, this Pheasant is quite common, and I have here seen in mid-July little chicks only a few days old. Around Wa shan this bird is fairly common, but from all accounts it is much more abundant in Yunnan. The habitat of this Pheasant is similar to that of Holderer's kind, and it affords similar sport.

Around Sungpan, in the north-west corner of Szechuan, occurs in quantity a Pheasant which closely resembles Anderson's, but is even smaller and rather different in colour. It may be a local form of this species. The predominant colour is a rich dark coppery bronze with dark green chest and breast, some feathers of the wing and rump are slate coloured ; head and neck very dark purplish-green, shading to black on the throat ; length, 28 to 30 inches, tip to tip ; average weight, 2½ lbs.

This Pheasant ranges up to the limits of cultivation (*circa* 11,500 feet), and is partial to brush-clad mountain-slopes bordering the fields of wheat and barley, the staple crops in this region. He descends the Upper Min Valley to about 6000 feet altitude, but is essentially a bird of high elevations.

At Sungpan I have shot this bird inside the city walls, and so abundant are they there that the Chinese declare that in winter they can be walked up to and killed with a stick ! Though this is probably an exaggeration the bird is undoubtedly very common, and the broad valleys and fairly easy slopes render the sport enjoyed after this Pheasant less fatiguing than that after any other of the mountain species. In the brushwood haunts of this bird, however, the Sallowthorn (*Hippophaë salicifolia*) abounds, and one's shins and knees need good protection against the stout thorns which beset this shrub. The true *Phasianus elegans* and this variety (or species, as the case may be) are hardy birds, and their introduction into the west is much to be desired. In North America in particular they would probably prove of greater value than the varieties of the more tender *P. torquatus*. They are as strong on the wing as any kind of Pheasant, lie close and afford the finest of shooting.

On the mountains bordering the Red Basin from Wênch'uan Hsien northward to the borders of Kansu, from 4000 to 9000 feet elevation, the Pheasant commonly met with is *P. berezovskyi*. In this bird the crown of the head is purplish ; neck, dark lustrous green ; chest and breast, rich coppery-bronze, with the breast-feathers narrowly margined with black ; flanks, dark ; rump, slaty-blue ; total length about 36 to 38 inches.

On one occasion, in 1908, two companions and myself enjoyed some excellent sport after this bird in the mountains to the immediate east of Mao Chou. Had we been out for a bag we could easily have secured a hundred brace in the four days we spent in this region. One of the cocks shot measured over 40 inches, tip to tip. Scrub-clad mountain-slopes near cultivation is the home of this handsome Pheasant.

Some confusion has arisen in the matter of identification of the Ringless Pheasant (*P. decollatus*) through l'Abbé David and others referring the Ringless Pheasants found in Shensi

and western Szechuan to this species. As far as I can determine this Pheasant is restricted to the low hilly country
bordering the Yangtsze River from Wan Hsien westward to
Sui Fu. It is also found in the valley of the Min River around
Kiating Fu, which district may be regarded as its northern
limits. It ranges from the river-level up to about 3000 feet
altitude, and possibly to 1000 feet higher south of Chungking,
which is about the regional optimum of the species. Scrub-
clad hillsides and thin woods bordering cultivation in the
Yangtsze region of the Red Basin is the home of this Pheasant.

The species was founded by Swinhoe (*Proceedings, Zoölogical
Society, London*, 1870, p. 135), on a bird purchased by his Boy
in the market at Chungking on 13th May 1869 ; a Pheasant
without a ring was a surprising novelty to Swinhoe, and he
could scarcely credit the story of a Ring-neck bird being unknown around Chungking.

This Pheasant is characterized by having the crown deep
brown, with the feathers margined with bronzed reflections ;
no white superciliary markings and no indication of white
neck-ring ; bare red patch on face very small ; entire neck
duck-green with purple reflections ; feathers of upper back
have the centres black with a narrow, medium, yellowish streak
and broad chestnut cross mark ; breast, chestnut-brown, with
broad black margins reflecting green ; flanks, buff ; tail,
broadly barred with black ; total length, 36 to 39 inches.

It is distinguished from *P. holdereri* by its differently coloured
breast, longer tail, and different markings on the feathers of
the upper back. Its closest ally is *P. berezovskyi* (*ante*, p. 112).
P. decollatus and its allied species form a well-marked group,
in which the greenish colour of the neck-feathers stops abruptly
at apex of the breast, forming a sharp line of demarcation in
colour. In *P. elegans* and allied species, which constitute
another well-marked group, no such line of demarcation is
found, the colour of the lower neck and chest merging gradually
into that of the breast.

Strauch's Pheasant (*Phasianus strauchi*) occurs in Kansu
province and on the mountains bordering the Amdo region,
frequenting woody places up to 10,000 feet altitude, and may
possibly extend into north-western Szechuan. It is described

as near *P. decollatus*, and is distinguished by the chest- and breast-feathers being narrowly margined with black ; flanks, darker ; mantle, fiery orange with narrow wedge-like apical streaks of blackish green, broad scapulars margined with dark maroon-red ; tail, more rufous-grey. The chest- and breast-feathers are bright, fiery chestnut-red, edged with purplish green ; flanks, bright chestnut-red, tipped with purplish green ; middle of breast and sides of belly dark green.

It is possible that the Pheasant found in the neighbourhood of Sungpan Ting should be referred to this species. Unfortunately my notes, made in 1904, are too incomplete to hazard an opinion either way. My impression, however, is that this Sungpan bird belongs to the *elegans* rather than the *decollatus* group.

Apart from species of the genus *Phasianus*, a large number of birds commonly spoken of as Pheasants in the broad sense of the term are found in the country with which we are concerned, and I now propose to deal with them in detail.

In the wooded country north and south of Ichang, between 2000 and 5000 feet altitude, the Reeves Pheasant (*Syrmaticus reevesi*) is abundant. This region is the real home of this magnificent bird. Westward he ranges as far as Lu Chou, but I never saw or heard of one west of the Min River ; northwards his range extends into southern Shensi. Every year numbers of badly prepared skins are brought into Ichang for sale. In Chungking dead birds are frequently to be seen on sale in the market. The flesh is very white and firm, but scarcely equal in flavour to that of a common Pheasant.

Marco Polo makes mention of this remarkable bird, and specimens were secured by Mr. Thomas Beale in Canton during 1808. Mr. John Reeves sent specimens to England in 1832. Nevertheless, it is only comparatively recently that its habitat has become known, and very few have seen the bird truly wild, and fewer still have shot it. Though I have seen many hundreds in their native woods, I have not shot more than a dozen. My largest specimen measured 6 feet and $\frac{1}{2}$ an inch. A bird shot by my associate, Mr. Zappey, in January 1909, and now in the Museum of Comparative Zoölogy, Harvard College, measures 6 feet 9$\frac{3}{4}$ inches, tip to tip. The largest

1. THE REEVES PHEASANT (SYRMATICUS REEVESI) ♂ 81¾ IN. ♀ 32 IN.

2. THE ANDERSON PHEASANT (PHASIANUS ELEGANS) ♂ 29 IN. ♀ 21 IN.

specimen I ever saw was shot near Nanto, at the head of the Ichang Gorge ; it measured 7 feet 2 inches, tip to tip !

The Reeves Pheasant is now so well known in aviaries that the following description is scarcely necessary : Crown and throat, white ; upper-parts, dull yellow ; feathers, narrowly margined with black, giving a scaly appearance ; breast, spotted and barred with black, white, and chestnut, on the sides the chestnut colouring shading into deep rufous-red; abdomen, black; tail, grey, barred with black to tip. The female averages about 32 inches, tip to tip, and is a very pretty bird ; the entire plumage is mottled, black, white, and brown, with the outer tail-feathers barred deep rufous-red.

Rocky, well-wooded country, where the undergrowth is not dense and in the neighbourhood of cultivation throughout the altitudes mentioned above, is the haunt of this bird. He has a partiality for oak woods and is very fond of acorns ; the pulpy fruit of various Rosaceous plants, especially of Coton-easter, is another favourite food.

The Reeves Pheasant is a wary bird and a great runner, quickly zigzagging to the mountain-top, from whence he prefers to take wing. He is very quick on the wing, shooting up through the trees at a sharp angle and then sailing from one ridge across to another. It is a fine sight to see this bird on a sunny day sailing across from ridge to ridge ; the great length detracts from the spread of the wings and he resembles some strange Chinese kite floating high up across the valleys. A strong bird, he flies with little apparent effort, and always puts at least one ridge between himself and the foe that caused his flight, and usually alights on a tree. The female when startled behaves similar to the male in running to the mountain-top. She then takes flight, making a curious chickering noise, and quickly dodges behind a tree-trunk. A common practice with the female bird is to alight on the upper branches of some convenient tree before essaying a long flight. The curious, weak, twittering call is more like that of some small animal than that of a bird.

The Chinese name for the Reeves Pheasant is " Ch'u che " (Arrow Chicken). The tail-feathers are largely used in Chinese theatricals. L'Abbé David suggests that this bird may be the original of the mythical Chinese Fung Hwang (Phœnix bird). To my mind this is extremely probable, but Williams in *The Middle Kingdom* considers the Argus Pheasant, found in Tonking and southern Yunnan, the origin of this fabulous bird.

THE GOLDEN PHEASANT

This well-known bird (*Chrysolophus pictus*) is abundant on the mountains of western Hupeh and eastern Szechuan, where it has much the same geographical range as the "Reeves." West of the Red Basin its place is taken by its congener, the Amherst Pheasant. Though so common, and in spring and early summer heard calling on all sides, the Golden Pheasant is rarely seen. Large numbers are entrapped alive by the Chinese and sold as pets. Few foreigners have had the luck to shoot this bird on his native heath. I had one chance only in my travels and that a very easy one, which I missed with both barrels. This bird frequents dense woods, where Evergreen Oak, Holly, Rhododendron, and other broad-leaved evergreen trees occur ; woods of Pine and Oak scrub are also a favourite haunt, and a strong partiality to rocky ground is shown. It ranges from 2000 to 8000 feet altitude, but is commonest between altitudes of 3000 to 5000 feet. This Pheasant feeds largely on berries, but is not averse to small acorns. He is a timid and crafty bird and seldom strays far from thick cover. He is also a great runner, only taking wing when hard pressed ; the flight is always low, fairly straight, and of short duration into the nearest thicket.

In the adult male the crest and rump are golden yellow ; feathers forming the cape, deep orange margined with black ; breast, flanks, and upper tail coverts tipped with scarlet ; tail, dark brown, barred with black ; average length about 42 inches, tip to tip. The young males resemble the females more than the adult males, having the head and rump rufous-chestnut ; rest of the body brownish and barred. The female is considerably smaller than the male, measuring 24 to 26 inches, tip to tip ; the colour of the plumage is uniform buff-brown, barred.

The Chinese designate this bird " Chin che " (literally,

Golden Chicken). To write it savours of vandalism, but this bird is really excellent eating, though there is little of him. Two or three guns properly posted, with some trained beaters or Sussex spaniels, might enjoy good sport after Golden Pheasants in the regions given above, but the work would involve plenty of hard climbing.

LADY AMHERST'S PHEASANT

West of the Red Basin this Pheasant (*Chrysolophus amherstiæ*) takes the place of the Golden Pheasant. The exact boundary line between the two species is difficult to determine, but I have not seen or heard of them being found in the same region. On Wa shan and Mount Omei and the jungle-clad regions west of these high mountains, the Amherst is abundant. North of Kuan Hsien he crosses the Min River, but the eastern limit everywhere is the western edge of the Red Basin ; north of lat. 32° he quickly disappears.

The habits and haunts of this bird are similar to those of the Golden Pheasant; the altitudinal range is from 3000 feet to about 10,000 feet in the south-west, and 8000 feet in the north-west of Szechuan. In these regions dwarf-growing Bamboos are a feature of the vegetation, forming absolutely impenetrable jungle. Such is the natural home of this bird, and though in season he is heard calling on all sides he is seldom actually seen. The Amherst is a very noisy bird, with a call very like the Golden. In the dense thickets it is of course impossible to shoot this bird, but in the early morning and late afternoon he is to be found in cultivated areas bordering the thickets, and occasionally a lucky snap-shot rewards the sportsman. The flight is similar to that of the Golden, and the natives entrap him in the same way. In the mountains bordering the Chiench'ang Valley in south-west Szechuan this bird must be very abundant, for the tail-feathers and cape are common articles of export from this region. They are used in Chinese theatricals in the same way as the tail-feathers of the Reeves Pheasant. The cape is also used in west Szechuan to adorn the caps of favourite male children.

In the adult male the crown, upper back, and breast are

resplendent dark green; rump, ochre-yellow with scarlet feathers in the upper coverts and on the lower rump; from the back of the crown projects a crest composed of a few long crimson feathers; the cape is white, margined with black, with outer feathers deep brown, barred with black; tail white, speckled and barred with black; length about 44 inches, tip to tip. The female is considerably smaller than the male and shows no sign of the cape; the crown and hind neck plumage is washed with greyish; back, buff-brown, barred; chest, buff, with under-parts lighter.

The Amherst has long been known in Occidental aviaries, and some interesting crosses between it and the Golden have been made. To my mind this bird is the most beautiful of all the Pheasants found in Western China. A colloquial name around Wa shan for it is " Kwong-kwong che." The shooting of this Pheasant, save by chance, is very difficult, but there are places where, by adopting the methods advocated for securing the Golden Pheasant, a few birds at any rate would reward an ardent sportsman. As a table-bird the Amherst is scarcely worthy of consideration; the flesh is coarse and without flavour.

BLOOD PHEASANTS

A common bird in the upland thickets between 8000 to 12,000 feet elevation throughout western Szechuan is *Ithagenes geoffroyi*. Around Tachienlu it is abundant, especially in thickets of Evergreen Oak and Juniper. This bird lies very close and is usually found in small coveys. When pressed by the dog it flies up into the taller bushes, making at the same time considerable noise, half fear, half scold in tone.

In the male the crest is dark grey; feathers of the back and chest lance-shaped, grey, each with a fine longitudinal white stripe; breast and sides, light green; under tail-coverts and few of upper coverts, crimson; tail-feathers, light grey, edged with crimson; spurs, 1 to 4 on each leg; length, about 18 inches, tip to tip. Female very similar to male; length, 16½ inches, tip to tip.

On Wa shan Mr. Zappey shot specimens of what proved

to be a species new to science, and Messrs. Thayer and Bangs have done me the honour of naming it *I. wilsoni*. This new Blood Pheasant differs in its smaller size, measuring only 14½ inches, tip to tip. The wing is very much shorter, and the whole bird is only about two-thirds the size of *I. geoffroyi*. The colour of the plumage is similar in both species.

In a region considerably to the west of Tachienlu (Yerkalo, on the Upper Mekong River), the Himalayan species (*I. cruentus*) has been reported, but I have no personal knowledge of this bird.

The colloquial name throughout west Szechuan for the Blood Pheasants is " Song che," which may be interpreted " Chicken of the thickets." This bird feeds on Juniper berries and buds of Larch amongst other things, and the whole flesh is permeated with a decided flavour of resin, rendering it unfit for the table.

PUCRAS PHEASANTS

Three distinct species of Pucrasia are now known from China, two of them occurring in the regions with which we are concerned. In western Szechuan, ranging (at least) from Wa shan in the south to Tachienlu in the west, and north-wards to Kansu, *Pucrasia xanthospila* is met with. This is essentially a woodland bird, frequenting the forests of Spruce and Silver Fir between 8000 feet and the tree-limit (11,000 to 13,000 feet, *circa*), where the undergrowth is mostly composed of Rhododendrons. It is particularly partial to places where fir needles cover the outcropping rocks. In such places in the forests these birds are frequently to be seen walking silently about with the dignified deliberateness of a barnyard fowl. They are silent (almost uncannily so) in their movements ; they skulk about amongst the timber, and refuse to take wing unless very hard pressed by a dog, when they fly up into the branches of the nearest tree. The males measure 22 to 23 inches, tip to tip ; ground colour of sides and flanks, grey ; nape, rufous-yellow ; basal parts of outer tail-feathers, grey. The common name of this pheasant is " Sung che," literally, " Pine Chicken." (*Sung*, strictly speaking, denotes the genus Pinus only, but in Western China the term has a wider application, and includes Spruce, Silver Fir, and Larch, as well as Pines proper.) This species of Pucras Pheasant has a very wide distribution, extending through the mountain ranges of northern China to eastern Manchuria. It is everywhere esteemed as a table-bird, the flesh having a particularly delicate flavour.

A variety of this species was obtained in the Shensi province by l'Abbé David and named *P. xanthospila*, var. *ruficollis*. This is distinguished in having the side of the neck very deep red ; lateral white spot little developed and surrounded on all sides by the metallic black ; median chestnut band less extended on the belly than in the type; black tints more developed on the back and wings. Very probably this should rank as a distinct species, but more material is wanted to determine this point.

In western Hupeh a new species has recently been reported and named *Pucrasia styani*. This bird measures about 18 inches, tip to tip ; the middle of the chest, breast, and under-parts are streaked like the sides, and there is no trace of the uniform chestnut band down the middle of the under-parts, which is characteristic of all the other species. The female is alike in all the species of Pucrasia, being similar in size and appearance to a common hen pheasant with a short tail and red legs. Styan's Pucrasia occurs in the vicinity of the Yangtsze River from near Kui Chou in Hupeh, westward (at least) as far as Yunyang Hsien, in eastern Szechuan. I flushed a small covey near Kui Chou in February 1901 and secured a female. The birds rose after the manner of ordinary hen pheasants, but scattered in several directions. Near Yunyang Hsien I saw several others in more open rocky ground. Stony, brush, and Pine-clad hillsides of no great altitude appear to be the home of this rare and interesting bird. As to how far it ranges to the north and south of the Yangtsze River is not known.

TEMMINCK'S TRAGOPAN

This strikingly handsome bird (*Tragopan temmincki*) is fairly common in parts of western Hupeh and western Szechuan between 4000 and 9000 feet altitude, frequenting woods and shrub-clad country. It prefers steep mountain-slopes, covered with arborescent vegetation, and in summer, when the foliage is on the trees, is most difficult to find. In winter it may occasionally be surprised, early in the morning and evening, near the margins of cultivation and close to thick cover. Like all the woodland pheasants these birds will only take wing when hard pressed and usually afford only a chance snap-shot. A heavy bird, the Tragopan flies at almost the speed of an ordinary pheasant, and always makes straight for dense brush or timber. The Chinese entrap them alive in the same way as they do the Golden and Amherst Pheasants. They are esteemed highly as pets and they sell for 3 to 5 ounces of silver each—a high price in these regions. The markings on the wattle are supposed to resemble the Chinese character for longevity, hence the common name, " T'so che." They are regarded as birds of good omen, bringing good-luck and long life to their fortunate owners. Every year numbers are brought down to Ichang for sale, where they find ready purchasers. In the mountains they apparently adapt themselves to captivity, but in the Yangtsze Valley proper the climate is too hot for them.

In the male the plumage of the upper-parts of the body generally is dark brownish-crimson, spotted with small whitish spots ; breast, indian-red, blotched with grey ; crown, rufous-crimson ; ears and lower eye-patch, indigo blue ; wattle, indigo blue with flesh-coloured markings ; tail, short and broad ; total length, 24 to 26 inches. The female has no wattle and the general colour of the plumage is brownish-buff, barred and spotted with darker colours ; total length, 18 to 20 inches.

The short tail and heavy body make the birds appear heavy in flight, and shooting them would be moderately easy did one but get fair chances. The Tragopan is a good table-bird, but to shoot them for this purpose alone would be gross vandalism. They feed on grain and berries, and are especially fond of the fruits of Cotoneaster and allied shrubs and of maize. South of Ichang this bird is much rarer than in the mountains north-west of this town and in western Szechuan.

The other species of Tragopan (*T. caboti*) found in China is confined to the eastern part of the country, being found in the provinces of Fokien and Kiangsu.

EARED PHEASANTS

Of the three species of these birds (*Crossoptilun*) known from China two are found in the far west. The only one I have seen and shot is the " White " or " Thibetan " species (*C. tibetanum*), which is abundant in the neighbourhood of Tachienlu. This bird frequents the upper timber-belt between 9500 and 13,000 feet, being commonly met with in large flocks, more especially in autumn, when it is probable that several coveys join forces. West of Tachienlu on the highway to Batang it is frequently to be seen strolling about in open grassy places and across the roadway. The walk is suggestive of a fine farmyard rooster, and with its broad, slightly raised, arching, plume-like tail the bird looks very stately. It is a great runner and always makes straight up the mountain-side into thick cover. When flushed it takes wing with the speed of a bullet, and with its heavy body makes a great noise on rising. The flight is of short duration and only attempted as a last resource ; generally the bird alights on trees.

The male has the crown black ; wing primaries, blue-black ; secondaries, blue-black shading into ashy-grey ; upper wing-coverts, grey ; middle tail-feathers, ashy-grey shading to iridescent blue-black with green reflections ; rump and abdomen clear white, rest of the plumage creamy white ; feet, coral-red, and the legs armed with long murderous-looking spurs ; total length, 38 to 40 inches. The female is similar to the male, with a total length of 34 to 35 inches.

Hunting this strong-legged, handsome bird is most " winding " and fatiguing sport. A favourite food is wild onions, and the strong flavour of this esculent permeates the flesh, which is dark-coloured and coarse and of little value for the table. The average weight of adult male is about 8 to 9 lbs.

This Crossoptilun ranges throughout the sub-alpine regions, bordering the timber-line from south-west of Tachienlu to the neighbourhood of Sungpan Ting and is one of the commonest birds found in this region. The vernacular Chinese name for this bird is " Ma che " ; a Thibetan name is " Shar har." How far to the south and west of the regions indicated this bird ranges I have no knowledge.

1   2

1. THE THIBETAN EARED-PHEASANT (CROSSOPTILUN TIBETANUM) ♂ 38½ IN.
♀ 34½ IN.

2. THE ICHANG PHEASANT (PHASIANUS HOLDERERI ♂ 34½ IN. ♀ 23½ IN.

The eggs are described by Pratt as being "light olive-dun colour."[1] Brooding commences about beginning of June and possibly earlier. By the end of July the "chicks" are of good size and strong on the wing.

According to l'Abbé David,[2] the "Blue-Eared Pheasant" (*Crossoptilun auritum*) occurs in the north-west of Szechuan and extends northwards to the Kokonor region, but is everywhere rare. He also says it is called "Ma che" (Ma ky), a name cited above as applied to the Thibetan-Eared Pheasant. I have no personal knowledge of this Blue Crossoptilun, but in the neighbourhood of Sungpan I was informed that "Ma che" occur, but are rare. I had presumed the white kind was meant, since the vernacular name was the same, but very probably I was mistaken.

In size and shape the Blue Crossoptilun is described as being similar to the Thibetan species; the ear-tufts are longer; body, slate-blue; tail-coverts passing from slate-blue to metallic black, lateral tail-coverts pure white in basal half; under-throat, white; eye-patch, blood-red; feet, coral-red. Female similar in plumage to male, but slightly smaller in size.

The specimens David sent to Europe he secured in Peking and I can find no record of any specimens having been shot in north-west Szechuan by a foreigner. Future travellers will do well to investigate this bird more fully, for there is a possibility of the species being distinct.

MONAL PHEASANT

Scattered through the same region as the White Crossoptilun, only at greater altitudes, occurs the magnificent "Monal Pheasant" (*Lophophorus lhuysi*), at once the most gorgeous and rarest of all game-birds found in these regions. Both David and Pratt comment on the rarity of this bird, and my experience is in accord with theirs. The King of Chiala detailed hunters specially for the purpose of securing specimens for Zappey, but no birds could be found. I was informed this bird was comparatively common east-north-east of Sungpan Ting,

[1] *The Snows of Thibet*, p. 202.
[2] *Les Oiseaux de la Chine*, i. p. 406.

in rocky places between 13,500 and 14,500 feet altitude, but I never met with one in that region. The only specimen that came under my observation was strolling about the margin of rocky scrub immediately above a wood of alpine Larch on the Ta-p'ao shan (between Romi Chango and Tachienlu), alt. 12,000 feet. In this particular locality I was told the Monal was fairly plentiful, but I doubt it. Hunters are ever on the look out to shoot and trap this bird, and the species is undoubtedly threatened with extinction.

In the adult male the top and side of head is metallic green with violet reflections ; eye-patch naked, very bright blue ; occipital tuft of long feathers, purple with metallic reflections ; back of neck and upper part of back intense golden-copper colour ; upper side of wings with bright blue and green reflections, washed with golden green on the shoulders ; lower part of back and rump white, with some angular blue spots on side of upper tail-feathers, the longest of which are steel-blue ; underparts of body black, glossed with green ; tail rather broad and rounded ; coverts, black and green with white spots ; legs feathered to the spur, which is stout ; below spur the legs and also the feet are greenish-brown ; total length, 36 to 40 inches. The females are brown, mixed with blackish and grey. The males assume adult plumage the second year ; in their first year's plumage they are similar to the female birds.

The magnificent bird has several local names. Around Tachienlu it is commonly called " Hwa-t'an che " (" Oak Charcoal Chicken "), or " Hoa-t'an che " (" Burning Charcoal Chicken "), both names having reference to the colour of the upper-part of the back and neck, which resembles the intense glow of a charcoal fire in full blast. A Thibetan name, which is used around Tachienlu and Sungpan Ting, is " Koā-loŏng." This name has reference to, and indeed simulates, the call of these birds, which is clear and distinctly quadrisyllabic. This call is usually heard in the early morning, but in wet weather it may be heard at any time of the day.

A favourite food of this bird is said to be the bulbs of various species of Fritillaria. The bulbs, known as " Pei-mu," are highly valued as medicine by the Chinese, and many men earn their livelihood collecting these and other medicinal herbs in

1   2

1. LADY AMHERST PHEASANT (CHRYSOLOPHUS AMHERSTIAE) ♂ 44 IN.

2. A PHEASANT GROUSE (TETRAOPHASIS SZECHENYI) ♀ 18 IN.

the alpine regions of Western China. According to l'Abbé David, a local name for the Monal is " Pei-mu che " (Paé-mou ky), in consequence of its feeding on these bulbs. Around Tach-ienlu this name is applied to a Pheasant Grouse described below, but it is highly probable that both birds are sometimes known by the same vernacular name. Mr. A. E. Pratt (loc. cit. p. 203) reports that he succeeded in introducing a single specimen of the magnificent Monal to England and handing it over to the Zoölogical Society, together with several *Crossoptilun tibetanum*.

In the province of Kweichou the mountains do not approach the snow-line, and in consequence it seems highly probably that l'Abbé David (loc. cit. p. 404) was wrongly informed as to this bird being found there.

PHEASANT GROUSE

This fine bird (*Tetraophasis szechenyi*), commonly called " Pei-mu che " by hunters around Tachienlu, is a denizen of the alpine woodlands between 12,500 to 14,000 feet elevation. West of Tachienlu, towards Litang, and more especially on the slopes of the Rama-lal Pass, it is fairly common, but always in open timber near the upper limits of the forests. It takes wing with the characteristic grouse whirr and swings through the glades at a great speed. The plumage of the adult male is : wings brownish, feathers margined with whitish buff ; throat, chin, and forepart of the neck pale fawn colour ; breast, slate with triangular black spots ; rump, light grey ; tail, greyish-brown, tipped with broad band of white ; wattle side of head, orange-scarlet ; total length, 18½ to 19 inches. Female similar to male, but about half an inch shorter. This is a very heavy bird for its size and most excellent eating. Mr. Zappey, who shot quite a number, considers this the finest for the table of all gallinaceous birds found in Western China.

In Mupin, a small principality a little to the east-north-east of Tachienlu, l'Abbé David secured the type of this genus, *Tetraophasis obscurus*. This species is distinguished from that described above in having the chin, throat, and forepart of the neck dark chestnut colour. In size the two species are very similar. David says the local name is " Yang-ko che,"

which may be translated the " Chicken of the Western kingdoms," signifying that it is peculiar to the Chino-Thibetan
borderland.

THIBETAN HAZEL-HEN

This interesting bird (*Tetrastes severtzovi*) is fairly common
throughout the Chino-Thibetan borderland, where it frequents
the upper timber-belt, bordering the brush-clad moorlands.
Alpine woods of Larch, such as occur on the Ta-p'ao shan,
north of Tachienlu, is a favourite haunt of this bird. It is a
plump, rather small bird, fairly easy to shoot, and of excellent
flavour. In the male the upper-parts are brown, mottled
with black; throat, black; under-parts, whitish-grey; chest,
mottled; total length about 14½ inches. Female similarly
coloured to the male, but without any black on the throat;
total length, 12½ inches. The dark, rich-brown general
colouring, darker and more distinct markings on the breast
and abdomen readily distinguish this bird from its near ally
the " Hazel-hen " of northern Europe.

SNOW-COCK

This denizen of upper alpine moorlands is rather rare and
very difficult to shoot, too difficult for me, in fact, and
though I have seen several I never succeeded in bagging a
specimen. At 14,000 to 16,000 one has very little breath to
spare, and a strong bird which prefers running and hiding
among rocks to flight has considerably the best of the contest.
As known to me, this bird (*Tetraogallus henrici*) is solitary
or in pairs, with a penchant for hiding amongst the boulders
of old moraines. The plumage of the back is grey, finely
vermiculated with pale buff; wing-coverts with large buff
and pale chestnut spots; crown, ashy-grey with chestnut and
white markings on side of head; throat, white; chest, grey;
abdomen, dark grey, striped chestnut on the sides; total length,
20 to 22 inches. This bird is very solid, with a heavy, almost
vulture-like, beak.

The only place I have seen this bird in is the neighbourhood

1   2

1. A BLOOD PHEASANT (ITHAGENES GEOFFROYI) ♂ 18 IN. ♀ 16½ IN.

2. THE THIBETAN HAZEL-HEN (TETRASTES SEVERTZOVI) ♂ 14½ IN. ♀ 12½ IN.

of Tachienlu, and it was here that Prince Henri d'Orleans secured the type-specimen. As far as I know no specimen of Snow-cock other than Prince Henri's has been collected in this region and sent to Western museums. L'Abbé David and others have stated that the Himalayan species (*T. tibetanus*) occurs in this same region, but no specimens have been taken. It is improbable that two species so closely allied inhabit the same locality. The principal difference is the grey chest of *T. henrici*, and white chest, divided from breast by a grey band, in *T. tibetanus*. Personally, I am of the opinion that the only Snow-cock found in the neighbourhood of Tachienlu and Mupin is *T. henrici*.

SNOW-PARTRIDGE

Colloquially known as the Hsueh che (literally, " Snow Chicken "), this bird (*Lerwa lerwa*) is not uncommon in the alpine moorlands of the Chino-Thibetan borderland. I have met with it around Sungpan Ting and on the Pan-lan shan, between Kuan Hsien and Monking Ting, at elevations of 11,500 to 14,000 feet. Hereabouts it is found on open moorlands amongst herbs and dwarf brush, always in small coveys of six to ten birds. They lie very close, and when they do take wing scatter in all directions. If flushed on the slopes of the mountains the birds fly down and round at great speed and are difficult to shoot. They make considerable noise when rising and never fly any great distance. Good sport can be enjoyed behind a well-trained dog, and the bird is very good eating. The plumage of the back is barred black with yellowish-grey and buff; under-parts, chestnut with a few pale stripes; beak and legs, coral-red; total length about 14 inches.

SIFAN-PARTRIDGE

This dainty bird (*Perdix hodgsoniæ sifanica*), which is about the size of the European Partridge, is common on scrub-clad mountains from Tachienlu to Sungpan Ting at elevations between 9000 and 14,000 feet. It is generally found two or four together, and in late August and September in small coveys of ten or twelve birds.   They lie close, and when flushed scatter and fly low and straight down and around the mountain-sides at great speed.   The plumage is brownish and barred all over, with a distinguishing chestnut collar : total length (male) about 11½ inches ; (female) about 10½ inches. This bird affords similar sport to the Snow-partridge, and is also excellent eating.

BAMBOO-PARTRIDGE

This bird (*Bambusicola thoracica*) is very common in western Szechuan up to 2000 feet elevation, but I never met with it in the eastern part of the province nor in Hupeh.   It is commonly found in clumps of Bamboo around houses, more rarely in dense scrub and margins of copses.   A ditch over-grown with rank weeds and shrubs is its favourite haunt.   It is usual to find coveys of ten to twelve birds ; they lie very close, and will not take wing until hard pressed by the dog.   They skulk and run, and, when forced, rise in a " bunch " with much noise, and scatter in all directions.   They do not travel far on the wing, and usually two or three merely fly up into the nearest bush ; it is seldom, too, that the whole covey takes flight, one or two stragglers generally skulk behind.   The bird is swift on the wing, flying low and straight to the nearest Bamboo clump or thicket.   In consequence of this and the fact of their usually being found near houses, the shooting is highly dangerous.   It is always snap-shooting, and people are every-where around, so that the sport is tantalizing at best.

This bird is a rather gross feeder, and might almost be termed a scavenger.   The flesh is white, but often strong flavoured.   When found some distance removed from houses where they have fattened on sweet potato, pulse, and grass seeds, they are really good eating.   Around Kiating this bird is very common.   The throat and breast is bright chestnut with a grey crescent across the chest ; crown, grey ; back, greyish-olive with chestnut markings ; wings with pale greyish marking ; belly, buff ; sides spotted with very dark chestnut ; tail, dull chestnut with pale vermiculations ; total length, 11 to 12 inches.   Female similar to male.

1

2

1. THE SIFAN PARTRIDGE (PERDIX HODGSONIAE SIFANICA) ♂ 11½ IN. ♀ 10¾ IN.

2. A BAMBOO PARTRIDGE (BAMBUSICOLA THORACICA) ♂ 12 IN.

Though a quiet-coloured bird, the Bamboo-partridge is really very handsome, and the colours of its plumage harmonize together splendidly. The males are great fighters. The Chinese commonly keep them in cages as pets, and derive much amusement from their pugnacious habits. The common name of these birds is " Chu che " (Bamboo Chicken).

Between the Ichang Gorge and the Niukan Gorge on the Yangtsze River Bamboo groves are a special feature. In these groves I have several times flushed an odd covey of Bamboo-partridge. This kind is smaller than the one described above, and is either a new species or *B. fytchi*, a species known to occur in western Yunnan. Unfortunately, our expedition did not secure any specimens, and I have no precise data.

WOODCOCK

The common Woodcock (*Scolopax rusticula*) is found scattered all over Western China. Anywhere and everywhere it may be found, but never in any great quantity. From October to the end of April, Woodcock are " in," and an odd bird is almost sure to be sighted in a walk after Pheasant or Bamboo-partridge. I have met with this bird from river-level (Ichang 120 feet altitude) up to 7000 feet altitude in western Szechuan in a variety of places. A favourite haunt is the side of a ditch, where there is a little cover. In spring, Woodcock are commonly to be found in the beanfields (Broad Bean, *Vicia Faba*), especially if there are a few trees near by to afford greater shade. Near Ichang in April 1907 I shot five within an hour, in a patch of beans beneath Plum and Pear trees, with houses not 50 yards away. When found in these moist shady beanfields the birds are usually very fat. One of the five birds alluded to above turned the scale at 15 ounces, and was a male at that ! It is commonly supposed that the female is larger than the male, but after measuring and weighing many birds I can find no decided difference. When the feeding is good the sexes attain about equal weight. The largest bird I have shot was a male.

As all who have given any attention to the matter know, Woodcock are to be found in the same spots year after year.

This is equally true in China as elsewhere. Though found with greater frequency in the vicinity of habitations where cultivation obtains and food in consequence more abundant, Woodcock are commonly met with on the mountains where little cultivation is carried on. When shooting east of Mao Chou in October 1908 at about 6000 feet elevation Woodcock were fairly abundant. I have also enjoyed good sport immediately outside of the city wall of Kiating Fu. The usual flight of a Woodcock is slow, rather erratic and owl-like, but when fairly roused there are few birds that fly at greater speed. All its movements in rising and flighting are silent, almost uncannily so.

PAINTED-SNIPE

This bird (*Rostratula capensis*) is more a Woodcock than a real Snipe, and is easily recognized by its curved bill. In plumage there is a vast difference between the young and adult birds. This difference is commonly attributed to sex, the better coloured birds being regarded as females. This is wrong. The adult birds are alike in both sexes. The primary quills of the wing are marked with buff-coloured, eye-like spots ; neck, deep chestnut shading to black on the breast ; outermost of the inner secondaries, white, forming a conspicuous stripe ; tail, olive-grey with four or five buff spots on both webs of the feathers, all of which are tipped with buff ; lower breast, white, this area passing on to the shoulder forming a stripe on the scapular region. The young birds have a much lighter plumage all over, and look very different, but a series will show every gradation up to the adult plumage.

The Painted-snipe has an exceedingly wide range, but I have only met with it in the neighbourhood of Ichang, where it arrives in September and remains to about the end of October. Some years it is more plentiful than others, but it is a rare bird at any time in this region, and I never saw it in western Szechuan. It is said to breed in the Yangtsze Valley, and I assume this refers to the alluvial reed-clad marshlands of the Yangtsze delta. At Ichang it is simply a visitant.

The favourite haunt of this bird is wet, weedy places,

including Lotus and other ponds where the Rush and False Rice (*Zizania*) are cultivated. The flight is low, similar to that of a Woodcock, affording easy shooting. Painted-snipe measure about 10 inches, and though very beautiful are of inferior flavour, and not worth shooting for the table.

SNIPES

Central and Western China has little to offer in the way of good snipe-shooting, and the phenomenal bags annually made in the Yangtsze Valley from Shasi eastward are not obtainable farther west. The high barrier mountains (Tsing-ling and Kiutiao ranges) running eastward from the Thibetan frontier and disappearing about long. 112° 30′ E. have probably more to do with this than anything else, the migratory flight of the main body of the birds being east of these ranges. In Szechuan there is plenty of good snipe-ground but very few birds. Snipe are not partial to the red sandstone soil, which predominates in Szechuan, presumably because it does not afford the best feeding ground. But most of the rice belt in this province has been so long under cultivation that the soil has been changed to black mud. Particularly true is this of the Chengtu Plain, which I have been told Snipe never frequent. This is not correct. Snipe can occasionally be purchased during the season in Chengtu city. I have shot them in several places on the Chengtu Plain, and in one instance found the birds fairly common around Mei Chou. I have also shot them around Kiating Fu and Hungya Hsien. In a marsh around the base of Wa shan during November 1904 I enjoyed some excellent snipe-shooting. These few facts show that Snipe are scattered over western Szechuan generally though sparingly.

Around Ichang quite a number of Snipe are shot annually, but the advent of the railway has destroyed the best ground. This strip of country, only some 2 miles long, was very dear to the heart of every foreigner interested in shooting who sojourned in Ichang. Now this much-loved spot is given over to railway-sidings, workshops, etc., and no longer affords any sport to the would-be shooting man.

In the regions we write of the three species common to the greater part of China occur, namely, Winter Snipe, Pin-tailed and Swinhoe's Snipe.

The Winter or Common Snipe (*Gallinago gallinago*) begin to arrive early in October, and some at any rate remain throughout the winter, migrating northward early in April. It is essentially a marsh and mud-loving bird, and is generally to be found in wet rice-fields, more especially those recently ploughed up; in muddy ponds amongst the Lotus (*Nelumbium speciosum*); in wet grass-clad marshes, sides of ditches, etc. When in good condition it weighs 4 to $4\frac{1}{2}$ ounces, but when it first arrives it is usually very thin and weighs no more than 3 ounces. Compared with the two following species the Winter Snipe is rather lighter coloured, more slightly built bird with rather longer legs and bill ; tail composed of 14 normal feathers all of the same size.

The " Pin-tailed " or " Lesser Spring Snipe " (*Gallinago stenura*) arrives from the north earlier and passes northwards again later than does the Winter Snipe, and it does not winter in the Yangtsze Valley. Around Ichang the Pin-tailed begins to arrive from the north about 20th August, and by the 1st of October has passed southwards. In spring it begins to arrive from the south about 1st April, and by 20th May has passed northwards. How this bird (and the same remarks apply to Swinhoe's Snipe) gets through the whole business of breeding and maturing the plumage on the young in so short a time (three months at most) is a mystery. I have never shot (nor heard of others shooting) a Pin-tailed in immature plumage or one which was obviously a young bird. That the birds should be hatched and reach adult size and plumage in such a brief period of time is one of the many wonders associated with migratory bird-life. Every one has of course shot birds varying considerably in weight, but this is merely a condition due to abundance or scarcity of food. When the birds arrive first from the north they are usually in poor condition.

The Lesser Spring Snipe frequents much drier ground than does the Winter Snipe. In spring it is partial to fields of wheat, pulse, and poppy, and grassy places either dry or rather wet. In autumn the favourite haunt is the fields of cotton and the margins of fields of maize and millet. In short, this bird favours cultivated crop-clad areas which the Winter Snipe, on the contrary, avoids. The Pin-tailed is readily recognized by its tail, which normally consists of 26 feathers ; the 10 central feathers are ordinary in appearance, and these are flanked on either side by 8 short, narrow, stiff feathers, from the presence of which the bird derives its name. The plumage generally is slightly darker and the bird rather stouter built than the Winter Snipe, though the scales show very little difference between them.

Around Ichang "Swinhoe's" or the "Greater Spring Snipe" (*G. megala*) is about as numerous in season as the Pin-tailed. It frequents the same haunts and arrives and leaves about the same time. In 1907 our first bird of the spring migration was shot on 27th March ; of the autumn migration on 26th August. These dates indicate pretty closely its earliest arrival in the two seasons.

Swinhoe's Snipe is much the largest and finest flavoured of the three common snipe. Its flight is slower, and it is easily recognized by its size, rather shorter bill, and normally 20 tail-feathers, of which the central 8 are ordinary with 6 narrow, stiff feathers of nearly equal length on each side. The colour of the plumage is similar to that of the Pin-tailed ; length, 11 to 12 inches ; weight, 6 to 8 ounces. When in good condition no finer table bird exists than Swinhoe's Snipe. I never met with this bird in western Szechuan.

The Solitary Snipe (*G. solitaria*) is to be met with on rare occasions throughout central and Western China. It is essentially a mountain bird, being partial to long grass and thin shrubberies bordering the sides of mountain streams. In the winter of 1900-1 I shot one bird immediately behind the town of Ichang, but this is the only one I have seen in the immediate vicinity of the Yangtsze River. On the mountains several days' journey south of Ichang at 4000 feet altitude, and again at 6000 feet altitude, I have shot solitary specimens ; also in north-western Hupeh at 5500 feet altitude I have secured this bird. In western Szechuan, around Wa shan, 5600 feet altitude, and around Mao Chou, at 5000 to 6000 feet altitude, I have been fortunate enough to shoot this bird.

It is, however, everywhere rare as far as my knowledge goes.

When in good condition this bird weighs 8 to 10 ounces, and is most delicious eating. It is the largest of the Snipes, measuring 12 to 13 inches. In the upper parts the plumage is uniformly dark brown ; under-parts, lighter brown, with the feathers narrowly edged with white ; tail of 16 to 24 feathers, the central feathers are normal, and are flanked by 4 to 6 narrow, stiff feathers on either side.

Latham's Snipe (*G. australis*) and the Jack Snipe (*Limnocryptes gallinula*) have been reported from eastern China, but I have never met with either in central and Western China.

QUAIL

This dainty little bird (*Coturnix japonica*) is found scattered all over central and Western China from river-level up to 7000 feet altitude, but is nowhere really common in these regions. Throughout eastern China it is abundant. Probably those found in the central and western regions breed there, whereas in the eastern parts of China they are largely migrants. These birds frequent dry grassy places, and are partial to the edges of maize and bean fields amongst the grass and weeds ; they are also commonly to be found amongst the dry stubble in rice fields before they are flooded and ploughed. They fly low and straight, and afford pretty and easy shooting when the crops are all cut. But when the maize is standing, the sport is very dangerous. Quail make straight for the standing crop, and as often as not Chinese will be found working hidden or half-hidden amongst the culms.

The densely populated nature of all the agricultural parts of China detracts considerably from the pleasure of shooting thereabouts. The danger of lodging pellets in some unfortunate native is ever present in the mind of the sportsman when after low-flying birds like Quail and Bamboo-partridge. Accidents happen to the most careful of shots, and the sport afforded by these birds in such places is not worth the risk. In parts of eastern China it is said (and there is no reason to question the statement) that the natives deliberately place themselves in dangerous places for the purpose of obtaining money if stung by pellets. Further, they are said not to be above malingering in this matter if there is a possible chance of money being forthcoming. In the west they are less sophisticated, and I never heard of such a thing happening.

The little Bustard Quail (*Turnix blandfordi*) is also fairly common around Ichang. It is easily recognized by the absence of the hind-toe, its rather long slender bill, and bright rufous-yellow chest. It measures 6 to 7 inches tip to tip, being about the same size as the Common Quail, and its haunts and habits are similar.

Grilled with a rasher of bacon and served on toast, Quail forms a tit-bit, worthy of any table. They are not so easily spoilt in the cooking as Snipe and Woodcock.

Quail are pugnacious birds, and are frequently kept as pets by Chinese on this account. Quail-fighting is a pastime much enjoyed in certain parts of China.

DOVES AND PIGEONS

It remains now only to say a brief word about the various Doves and Pigeons of this region. Up to about 4000 feet altitude in the Bamboo clumps and trees surrounding villages and homesteads the Common Turtle Dove (*Turtur chinensis*) is everywhere abundant. This pretty bird is inferior eating, and unless one is hard up for meat there is no excuse for shooting it. In the thin woodlands, and ranging up to about 6000 feet or even higher in well-cultivated regions, the Greater Turtle Dove (*T. orientalis*) occurs, but is much less plentiful than the Common Turtle Dove. This is a very good table bird, perhaps the best of its family. The Pallid Turtle Dove (*T. decaocta*) is also found scattered through north-western Hupeh and eastern Szechuan, but is nowhere common. Around Ichang and westwards into eastern Szechuan the small Turtle Dove (*T. humilis*) [1] occurs as a late spring visitant, and breeds

---

[1] Ornithologists now put the Doves in several different genera, and the species referred to above are spoken of respectively as *Spilopelia chinensis*, *Turtur orientalis*, *Streptopelia decaocta*, *Onopopelia humilis*.

there. This small bird has a very distinct, hoarse croaking note, and is partial to the tallest trees around houses and cultivation.

Of Pigeons proper at least 6 species occur, but 2 only are really abundant. The Rock Pigeon (*Columba rupestris*) is found in quantity throughout the valleys of the Upper Min River from near Wênch'uan Hsien (alt. 3900 feet) to beyond Sungpan (up to 11,000 feet altitude), where steep cliffs abut on cultivated areas. It is equally common around Monkong Ting, in the valley of the Little Gold River (Hsaochin Ho), and around Romi Chango, situated on the Upper Tung River. It descends the valley of the Tung River to near Luting chiao, but is not plentiful in that neighbourhood. This bird is also common north and west of Tachienlu and in the valley of the Yalung River. Indeed, it is generally distributed throughout the whole Chino-Thibetan borderland from 4000 feet altitude to the limits of cultivation (*circa* 11,000 to 13,000 feet). Large flocks are to be seen on all sides perched on the cliffs, in the fields feeding, or circling around. This Pigeon breeds in the holes in the cliffs, and excellent shooting can be had wherever these birds occur. For the table, however, it is inferior to the Greater Turtle Dove. In the Upper Min Valley all the villages are walled, and ruined forts and guard-houses are met with on all sides. Associated with these places, and breeding therein, and also high up in the cliffs, occurs a species of Pigeon which I assume to be *C. intermedia*,[1] a species very closely allied to the Rock Pigeon of Europe (*C. livia*). This Pigeon is easily domesticated, and under the eaves of their houses the villagers and peasants fix crude bamboo baskets for this bird to nest in. It is regarded as a bird of good omen, and is reputed to shun the haunts of evil-doers ! In the Min Valley this Pigeon is in a state of more or less semi-domestication, and the birds exhibit very considerable variation in plumage, many being indistinguishable from ordinary tame pigeons. It may be that the Rock Pigeon (*C. rupestris*) mentioned above enters somewhat into the production of this semi-domesticated race. Both races occupy much the same territory, and both are met with in flocks of from 20 to 100 or more birds.

[1] This species is possibly the one from which the Chinese domesticated races of Pigeon have been evolved.

In alpine regions, from 10,000 to 14,000 feet elevation, in proximity to snowclad peaks, there is a species of Pigeon which may be termed the " Snow Pigeon." This bird (*C. leuconota*) is larger than either of the foregoing species, with much lighter-coloured plumage. I first noted this Pigeon on the slopes of the Ta-p'ao shan north of Tachienlu. It also occurs on the Cheto shan and other places west of Tachienlu. It is met with in flocks, but does not appear to be common.

A Green Pigeon, possibly *Sphenocercus apicicauda*, is occasionally met with in the Chino-Thibetan borderland, but is rare, and probably only a summer visitant. The long tail and beautiful plumage render this a strikingly handsome bird. I met with it once only, and that was around the hamlet of Mao-niu, situated about midway between Romi Chango and Tachienlu. This little village is surrounded on all sides by large forests, and a small flock of Green Pigeons was circling around high up out of gun range. L'Abbé David mentions the Green-winged Ground Dove (*Chalcophaps indica*) as occurring around Mupin. The same authority says that two other Himalayan pigeons—the Spotted Pigeon (*Columba hodgsoni*) and the Long-tailed Pigeon (*Macropygia tusalia*)—are also occasionally met with around Mupin. I have no knowledge of either of these birds.

In north-western Hupeh Mr. Zappey and I saw on one occasion a couple of pigeons that looked distinctly green in colour, but we were unable to obtain specimens. Probably these birds were *Crocopus phœnicopterus*.

The Chinese name for Pigeons and Doves alike is " Pan-chu." My followers gave the name " Lu (green) Pan-chu " to the *Sphenocercus*, but this is the only kind to which I ever heard a special vernacular name applied.

Pigeons are everywhere domestic pets with the Chinese, and pigeons' eggs enter very largely into a much-esteemed Chinese soup. Like many other of their favourite foods and medicines, pigeons' eggs are supposed to possess aphrodisiac properties. A common practice in Western China is to fix on the top of the pigeon's tail at its base a small, round, hollow piece of wood having a slit on one side, which produces a humming, whistling noise as the birds circle around in flight.

CHAPTER XII

SPORT IN WESTERN CHINA

WILD-FOWL : SHOOTING ON THE YA RIVER

WILD-FOWL in great variety abound all over China, and the West has its share, though a lesser one, it is true. In that great alluvial plain and swamp bordering the Tungting Lake in central China they occur in myriads during the winter season. The same is true of the Lower Yangtsze delta. Throughout the region of the Gorges wild-fowl are comparatively rare, for the simple reason that steep cliffs and deep water are not to their liking. Above Kuichou Fu they are more common, but not nearly as much so as farther west. On the lower reaches of the Min River, and its tributary the Ya, which unites with the Tung at Kiating Fu, they are very plentiful. Sandbars difficult of access and stony places near the rapids and races of the more shallow parts of the rivers are favourite daytime haunts. At night the farmers' wheat and pulse fields near the rivers are freely visited. The wild-fowl which frequent Western China in the winter season probably breed in the Kokonor region, whereas those which visit the eastern parts of China breed in the tundras of Eastern Siberia. The mountains of western Hupeh, eastern Szechuan, and Shensi constitute barrier - ranges demarking the lines of migratory flight. Apropos of this boundary it is worthy of note that Geese have never been shot, neither have they been observed resting, west of Ichang as far as records and my own observations go. Yet to the east of this point they are probably more abundant than any other family of wild-fowl.

In the more eastern parts of the Empire, Chinese wild-fowlers find a lucrative business in supplying the markets of Shanghai and other large Treaty ports. They frequently employ methods peculiarly their own, and the following account, by a Chinese sportsman, is taken from that interesting book by H. T. Wade, *With Boat and Gun in the Yangtsze Valley*, pp. 139–41 :—

" *Catching Wild Ducks.*—At the close of a cold December, some 7 miles from the walled city of Kintang, near a large pond, I saw a man beckoning to me, and as I approached he asked me not to shoot the ducks in the pond. He explained that his friend was in the water ; so I waited to see what would happen. After some time his friend landed, wearing a large bamboo collar or cangue, and carrying a basket containing a few wild and three tame ducks secured together by a string. He was dressed in goat-skin, with the wool inside ; his stockings were stitched to the clothing, and so oiled as to be nearly waterproof. Thus accoutred, he immersed his body, using the cangue as a float. On his hat were placed bunches of grass, and on the cangue two or three decoy-ducks. He slowly approached the wild-fowl, and when near enough dexterously caught the unsuspecting duck by the leg, and dragged it under water. I watched him until he had gathered nearly the whole lot."

" *Shooting Wild Ducks.*—Probably no man in the world but the Chinese fowler would enter the water up to his neck, in the coldest weather, to shoot ducks. His *modus operandi* is like this : a light wooden frame or a small punt supports his gingal. The fowler lets the frame with its freight float in front of him, while he, following, is concealed from view by bunches of grass and weeds stuck into his hat. As soon as within range, which is invariably a very short one, he fires into ' the brown ' a heavy charge of iron shot. He never fires at two or three fowls, as his shot costs money. He bides his time, and then fires into the *brown*."

" *Catching Geese.*—A common method is to lay down a long line, to which is attached a number of thin bamboo slips, bent double, and the two ends of the bamboo inserted in a bean. This bait is laid on a regular feeding ground, and the hungry goose swallows it greedily, with the result that the act of swallowing liberates the bent bamboo, which, resuming its original shape, chokes the bird."

The annual slaughter of wild-fowl in China is enormous, but the birds are as wary as their kind is anywhere else in the world. On bright sunny days they are more easily caught "napping," but the man with a twelve-bore earns all the wild-fowl he bags in a season. At least, such is my experience. I have no intention of entering deeply into this subject, since the sport differs in no particular from that of the same nature elsewhere in the world. There is, however, a novel form of duck-shooting obtainable on the Ya River, in western Szechuan, which affords both excitement and good sport, and should not be missed by anyone to whom an opportunity offers.

The Ya, which unites with the Tung River a mile or so beyond the west gate of the city of Kiating, is a swift-running stream thickly bestrewn with boulders, shingle, and sandbanks. Boats are in use at the various ferries, but the river generally is unsuited to navigation, and merchandise is conveyed up and down on rafts. These latter are specially built for shallow rivers, and ply principally between Yachou Fu and Chong-peh-sha via the Ya River to Kiating Fu, thence on the Min to Sui Fu, and from there on the Yangtsze to Chong-peh-sha. With the exception of Lu Chou, Chong-peh-sha is the largest town between Chungking and Sui Fu, although it has no official status. It is a great and famous wine-mart, and rafts are very largely employed in carrying wine in large jars from this town to the inland markets situated on the shallow rivers farther west.

Although fragile-looking affairs, these rafts are quite unsinkable, and the best of their kind in existence. They are built entirely of the culms of a giant Bamboo, known as "Nan chu" (*Dendrocalamus giganteus*). Each raft is about 66 feet in length and 11 feet wide. The canes are laid side by side in one plane and securely lashed to numerous cross-beams, not a single nail being used in the whole construction. Several unequal lengths of bamboo are used so that the end-to-end joints occur at irregular intervals. The stern of the raft is square, the prow bent upwards to serve as a fender against rocks and shoals. The outer silicious "skin" of the canes is removed and the nodes hardened over a hot fire. The bending of the canes to form the upturned prow is done by

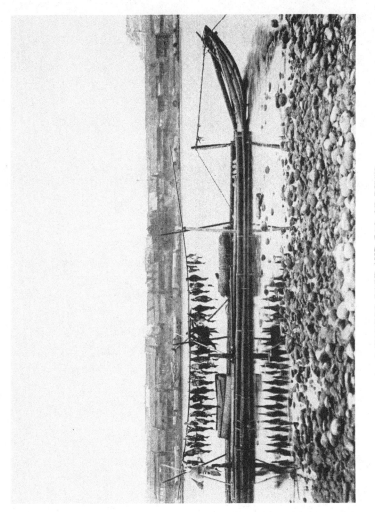

A BAMBOO RAFT AND BAG OF DUCK

heating and weighting with heavy stones. A narrow wicker-staging is carried down the centre of the raft, and is raised about a foot above the floor ; on this the merchandise is placed to keep it dry, or, in the case of wine-jars, they are lashed to the staging.

These rafts are capable of yielding both transversely and laterally, and can thus pass over any slightly submerged obstruction. Fully loaded, one raft will carry a freight of about 30,000 lbs. weight, and then draw only about 6 inches of water, owing to the great buoyancy of the hollow cylinders of bamboo. Down-stream a crew of four men manipulates each craft, which is propelled by an oar on either side and steered by a scull aft and another forward, but the latter is only used in the more difficult places. The sculls and oars are fitted to Alder stumps, which serve as rowlocks. The rafts are hauled up-stream by men attached to bamboo lines, and several usually travel in company, in order that the crews may assist one another over the more difficult rapids.

The Ya when not in flood is a clear-water stream, and from the raft the stony river-bottom is plainly visible ; often the boulders look so dangerously near the bottom of the raft that the passenger expects a bump every few minutes. A curious hissing and crackling noise accompanies the raft's progress over the more shallow places. This noise is due to the move-ment of the boulders and stones in the bed of the stream, the hollow bamboo tubes acting as sounding boards. There are many angry and dangerous rapids and whirlpools on the Ya, and the current is very swift : shooting these places is most exciting work. There is really very little possibility of an accident unless the raft is overladen, but as every rock and stone is visible in the clear water the uninitiated feel the presence of danger in a rather alarming fashion.

In the winter season this stony river is the haunt of thousands of Wild Duck, which congregate in the daytime in the vicinity of rapids, races, and boulder-strewn shoals. Ex-cellent and highly exhilarating sport may be obtained by engaging a raft at Yachou and shooting wild-fowl from it as the stream is descended. A little noise will scare the birds on the approach of the raft, and whilst the latter successfully shoots the rapid or race it is up to the man with the gun to bring down the ducks. The size of the bag depends largely upon the steadiness of nerve, but it takes a few cartridges before one can fairly well judge just how much to " lead " a bird when pulling the trigger. The movement of the craft, both forward and sideways, considerably increases the difficulties of aim. Two guns are best, one forward and the other aft. The dead birds are easily retrieved at the foot of the rapids ; the wounded ones are carried over by the force of the current, and can then be finished off. Those falling on land are difficult to mark down and retrieve. After a little practice a steady shot can make a good bag of duck from these rafts.

Early in December 1908 my companion, Mr. Zappey, accompanied me on a journey by raft from Yachou to Kiating, which occupied a couple of days. The weather was boisterous and wet, and wild-fowl comparatively scarce. We shot and retrieved 53 ducks, and probably killed in addition about a third of that number. Although the bag was not large the excitement and fun was immense. To anyone in search of exhilarating sport, Duck-shooting from a raft journeying down the Ya River can be confidently recommended.

The common wild ducks found in the west are Mallard, Wax-bill, and ordinary Teal. Others occurring there in lesser numbers are Falcate Teal, Spectacled Teal, Golden-eye, Pin-tail, Goosander, Smew, Pochard, Shoveller, Lesser Grebe, and Ruddy Sheldrake (Brahminy) (apropos of the latter it may be of interest to mention that I once found a couple breeding in the margin of an alpine lake near Tachienlu, at 15,500 feet altitude). Three species of Gull—two large grey kinds and a kittiwake—ascend to this region, 2000 miles inland from the coast. Widgeon I never saw in the west, and the same remark applies to the Mandarin Duck, Swan, and Geese. At Kiating the harsh cry of a very large kind of Crane may be heard any night during November, and on dull wet days small flocks may be seen flighting southward. Very seldom, however, do they alight in this neighbourhood, and still more rarely are they to be seen resting during the daytime. These birds winter around the lakes in Yunnan, and apparently make a post-haste thousand-mile flight thither from their breeding grounds in the Kokonor region.

The Goosander, Smew, Pochard, and one or two others are diving, fish-eating ducks, but if skinned they lose their fishy flavour and become palatable, but even then they are inferior eating in comparison with Mallard, Wax-bill, and the Common Teal.

CHAPTER XIII

SPORT IN WESTERN CHINA

RUMINANT AND OTHER GAME ANIMALS

I N the matter of large game animals Western China is of special interest, since the country being so little known there is a possibility of new and undescribed species or varieties rewarding the energetic sportsman-explorer. The difficulties in the way of any systematic exploration of the Chino-Thibetan borderland are to-day very great, and many years will elapse before the world is thoroughly informed on this fascinating region. My friend, Captain Malcolm M'Neill, of Oban, Scotland, visited this region in 1908, and in a brief season secured a nice collection of different trophies. These included the Takin and new varieties of a Bear and Stag, which Mr. R. Lydekker has named in his honour. Mr. Zappey whose primary object was the collecting of birds, found opportunity to shoot quite a number of animals, including the Takin ; the last-named, by the way, he was the first white man to secure by actually shooting the beast.

Quite a number of different kinds of game animals are now known from Western China, but only odd specimens of each have reached the Occident, and there is much yet to be learnt regarding every one of them. More especially is information needed on the habits, colour-variation, and geographical range of the different species and varieties. The affinity of the fauna is with that of Upper Burmah and the Himalayas as far as the animals of the forests are concerned, but in every instance peculiar species or subspecies obtain. The animals of the higher altitudes above the tree-line are mostly common to the whole of the Thibetan highlands. Indeed, the uplands of the Chino-Thibetan borderland constitute the eastern limit of the Central Asian fauna.

Personally I have hunted none of the larger animals, but I have been associated in the field with those who have. I have at different times seen in a living or dead state nearly all the animals described below, and in many ways have enjoyed exceptional opportunities for acquiring information. The following pages, compiled mainly from notes collected during my travels in this region, make no pretence of being exhaustive, but they may perhaps add something to the present scant store of knowledge.

BLUE SHEEP (BHARAL)

Bharal or Pan-(often pronounced Pai)-yang (*Ovis nahura*), as they are locally called, are common throughout the Chino-Thibetan borderland on the higher ranges above the timber-line. During the summer-time they frequent the alpine regions between 13,000 and 17,000 feet elevation. At Tachienlu (alt. 8400 feet) they have been shot in June on the cliffs that overlook the town itself. Around Sungpan Ting they are common, and in the uplands everywhere between the above points they are to be found roaming about in flocks, often of considerable size. In the winter they descend to 8000 or 10,000 feet altitude. When fired upon, the " Pan-yang " has a characteristic habit of running a short distance, then halting and looking round at the enemy.

The adult animal stands about a yard tall at the withers and has a long, narrow head, short ears, no mane or beard, and a thick, close coat of hair. The general colour of the upper-parts is brownish-grey tinged with slaty-blue, darker in summer than in winter ; under-parts white ; lower part of tail black. In adult rams the face and chest are black, with a black band along the flanks, white knee-patches, and a black stripe down the front of all four legs. The horns are blackish-olive with an S-like curvature, rounded and nearly smooth save for the annual rings of growth. The horns of the ewes are short, drawn together at the base, curving upwards and outwards in a somewhat scimitar-like fashion.

The " Pan-yang " is a rather heavily built animal, strong
and active, and very much at home amongst steep, difficult
cliff-country.   It is fairly easy to stalk, though the nature of
the country and the rarefied atmosphere render the work
tiring and arduous.   A full-grown animal weighs between
125 and 140 lbs.; the mutton is of good flavour, without
any suspicion of "goaty" odour.   The colloquial name, "Pan-
yang," is very descriptive, signifying " half-sheep, half-goat,"
thereby denoting the somewhat intermediate appearance and
character of this animal.   Shooting on the Hsueh-lung shan
range west of the Min River in the territory of Wassu on 13th
June 1908 Captain Malcolm M'Neill secured two heads, which
he informs me have horns measuring as follows :—

	Length	Circumference at base	Tip-to-tip
No. 1 . .	26¼ inches	12 inches	27 inches
No. 2 . .	25⅝ inches	12 inches	30 inches

N.B.—Extreme tip of the last head stripped, otherwise it
would probably have spanned more.

The Bharal of this region apparently differs in no particular from that found in the western parts of the Thibetan
plateaux.   Quite a number of these animals are annually shot
by hunters, and I have seen many skins on sale in the streets
of Chengtu city.   The skins are not much valued, being used
for lining the cheap winter garments worn by the lower
middle class of that region.

Another Sheep, probably Hodgson's (*O. ammon hodgsoni*),
occurs immediately to the west and north of Tachienlu, but is
very rare.   It has been seen in the neighbourhood of Litang
by at least two travellers, but there is no record of one being
shot.   This Sheep frequents the alpine regions above 13,000
feet.   Zappey saw three near the Rama-lal Pass, but failed to
get within comfortable range.   He says they were larger than
Bharal, as they should be if they are Hodgson's variety.

SEROW, " YEH LAU-TSZE "

There are two distinct-looking kinds of Serow in Western
China, but the colouring in these animals is variable, age having
much to do with it.   Zoölogists are not yet agreed upon the

systematic rank of these two varieties, but Mr. R. Lydekker (*Proceedings of the Zoölogical Society of London*, published April 1909), summing up all the evidence before him, considers them two distinct species. This arrangement seems logical, and is certainly more convenient than that of regarding them as forms of the Sumatran Serow (*Capricornis sumatrensis*), as some authorities do.

Throughout Western China this animal is common, and is everywhere known as " Yeh Lau-tsze " (Wild Donkey) or " Ai (Ngai) Lau-tsze " (Cliff Donkey), the long ears being responsible for the vernacular name. Between 5000 and 10,000 feet elevation in west Szechuan, Serow is probably the commonest wild animal. Around Wa shan, Tachienlu, Lungan Fu, and throughout the Upper Min Valley it occurs. I have seen specimens killed in all these places, and elsewhere also. The flat skins are commonly used as bed mattresses throughout these regions. Mr. Zappey shot two near Wa shan, and Captain M'Neill secured three specimens near Tachienlu. The lamented Mr. J. W. Brooke and his companion, Mr. C. H. Mears, shot a couple (at least) in the Upper Min Valley, below Wên-ch'uan Hsien. The above were all killed in 1908. But previous to this Messrs. Brown and Wilden, respectively of the British and French Consular services in China, had shot examples of this animal in the Upper Min Valley. In 1893-4 Mr. M. M. Berezovski secured specimens in the mountains north-west of Lungan Fu. The earliest known examples of these Serow were taken by l'Abbé David in the principality of Mupin in 1869. These animals are always found in wild, precipitous, brush-clad country, and, in consequence, are difficult to hunt. In the Upper Min Valley the mountains are mainly composed of mud shales, and landslips are frequent, rendering the hunting of these animals highly dangerous work. When startled, Serow plunge into the thickest cover on the cliffs, and are difficult to drive out into a position affording a decent shot. The natives snare them, hunt them with dogs, and shoot them, and occasionally capture them in dead-falls. The native dog is extremely useful in hunting Serow, commonly hounding them into positions where they cannot escape, save by rushing their tormentors. Though naturally timid, dogs madden them

into making wild rushes, and they are fierce and dangerous
when at bay. They have been known to kill the dog hunting
them and badly wound the hunters themselves. In steep,
difficult country an animal driven to charge by fear is
extremely dangerous owing to the precarious foothold
obtainable.

Around Wa shan the White-maned Serow (*Capricornis
argyrochætes*) is the common species; around Tachienlu
Milne-Edwards' Serow (*C. milne-edwardsi*) is the common
animal found. In the Upper Min Valley and around Lungan
Fu both species occupy the same regions, and this is probably
true for western Szechuan generally with one or other species
more common in certain districts.

A female of the White-maned Serow shot in May 1908
near Wa shan by Mr. Zappey gave the following measure-
ments : length , $66\frac{1}{16}$ inches; tail, $4\frac{3}{8}$ inches ; height at shoulder,
$35\frac{1}{2}$ inches. Colour : mane, light brownish; body and legs
darker and brighter than in Milne-Edwards' species (*infra*).

Zappey's experience was that the male and female kept
together. He shot a female in the later afternoon ; the dogs
remaining at the foot of the cliff all night kept the male in a
place from whence it could not escape, and Zappey returned
at daylight and shot it.

A male of Milne-Edwards' Serow, which Zappey secured
near Tachienlu, measured as follows : length, $66\frac{4}{8}$ inches ;
tail, $4\frac{4}{8}$ inches ; height at shoulder, 39 inches ; horns, $8\frac{1}{4}$ inches
long. Colour : mane, whitish, 10 inches long ; back of rump,
fore-legs to just above the knee, and hind-legs to half-way up the
thigh, chestnut ; back and sides, dark iron-grey ; belly, dark
grey.

The flesh of the Serow is dark coloured, tough, of poor
flavour, and the least desirable meat I have tasted.

On the high mountains of north-western Hupeh, forming
the Han-Yangtsze water-shed, a Serow occurs sparingly, and is
called " Ming-tsen Yang." The characters denoting this name
were interpreted by a Chinese gentleman as meaning " Clear-
maned Goat." This is a very appropriate name for the
White-maned Serow (*C. argyrochætes*), though it is possibly a
distinct animal. Neither Zappey nor myself succeeded in

obtaining one of these animals, though we came upon fresh tracks and dung in quantity. As a result of some ten days' hunting, Zappey only once sighted this Serow. A loud, angry snort in the brush, a momentary glance, and all was over. The animal covered 15 to 20 feet at a bound, and was through the thicket and over the ridge in less time than it takes to tell. As the country is everywhere difficult, and the animal scarce, there is very little chance of securing a trophy. On the high mountains south-west of Ichang this same Serow occurs, but is even more rare there than in the north-west of the province.

We secured fragments of a flat skin and several pairs of horns, but these are insufficient to show what the species is. A pair of these horns I obtained in exchange for a couple of empty bottles measure : length, 10¾ inches ; circumference at base, 5 inches ; tip interval, 4¼ inches.

The horns of all the Chinese Serow are very similar, being jet-black in colour, ringed, and tapering to a point ; the position is erect and curving backwards. The hair is coarse, long, and shaggy, with a short woolly underfur, and neither in the colour of the pelt nor in the size of the horns is there any marked difference between the sexes, age having more to do in these matters than anything else.

GORAL, " YEH YANG-TSZE "

Three species of Goral have been recognized in Western China, two in west Szechuan and one in west Hupeh. Quite recently they have been lumped under one species, but this is scarcely a satisfactory method of classifying them. The Goral found in Hupeh and Szechuan are readily distinguished by their colour, and it is convenient at any rate to keep them under separate names. These animals make their home amidst limestone crags and precipices, and though quite common are rarely seen, and not easily hunted. They are not so difficult to shoot as to retrieve afterwards, consequent upon the precipitous nature of their haunts. During sunny weather they lay up during the daytime on scrub-clad ledges of rock or in the mouths of caves that are so common in limestone regions. They feed in the early morning and in the evening, except during misty, rainy weather, when they are not particular.

The limestone crags and cliffs of the Yangtsze Gorges and the glens leading therefrom are favourite haunts of the Hupeh species. In the Ichang Gorge itself this animal is quite common, and anywhere in the limestone regions in western Hupeh up to 4000 feet altitude it is to be found.

The natives assert that Goral are found even up to 7000 feet altitude. There is a precipitous range near the tiny hamlet of Kuan-pao in Changyang Hsien, four days' journey south-west of Ichang, where they may be found. This range reaches 7000 feet altitude in its higher parts, and a variety of game occurs there. Quite a number of foreigners have enjoyed good sport after Goral in the glens around Ichang. Probably the first to shoot one hereabouts was Dr. Aldridge in the early 'eighties of last century. The late Père Heude named this animal *Kemas aldridgeanus* [1] in his *Les Memoires concernant l'Histoire Naturelle de l'Empire Chinois* (pub. 1880–1901). This naturalist made a great study of Eastern Asian mammals, and his specimens went chiefly to the Museum at Sicawei, just outside Shanghai. Unfortunately these specimens have been sadly neglected. The author above quoted was not sufficiently careful in the matter of defining his species, in publishing good descriptions, and in preserving his types. The consequence is, that much animal life has been wasted and the nomenclature of Chinese mammals rendered exceedingly difficult to later systematic zoölogists. Anyone who has seen the collections at Sicawei must regret that the types were not sent to Paris or some other western centre where they would have been properly looked after, and accessible for comparative study.

The Chinese who hunt Goral usually study their runs and snare them, or occasionally they shoot them. The method of hunting them for shooting is as follows : The man with his rifle (matchlock in case of natives) is posted on one side of a

---

[1] In accordance with modern nomenclature this name becomes *Næmorhedus aldridgeanus*. The specific name, *henryanus*, had been earlier applied by Sclater to this same animal. Milne-Edward's specific name, *griseus*, was published in 1871, and has priority over the above names if it be accepted that the Ichang and Mupin Goral represent the same species.

glen or ravine, high up at a point of vantage where a clear view may be had. The " beaters " then traverse the opposite side, hurling down rocks and making a great noise. This startles the animal, which then skulks along the ledges, through the brush until driven out on to bare cliff. If there is any possible ledge it will descend almost vertical cliffs, dropping easily 18 or 20 feet from ledge to ledge. Excellent rifle-shooting is afforded by this beast if only the beaters can be kept sufficiently out of the way of danger.

My companion, Mr. Zappey, enjoyed good sport after these Goral, and secured several specimens. The illustration (p. 152) shows an adult male and female and a young male shot by him in the western end of the Ichang Gorge during January 1908.

This same Goral is common to all the gorges, and in the long, gloomy, and forbidding Wushan Gorge it occurs in plenty. As an example of luck and good shooting I give the following experience :—

On our journey up river to west Szechuan in late March 1908 we were sailing up through the Wushan Gorge enjoying a moderately strong, fair wind, and were just off the hamlet of Nanmu-yuan. My companion, Mr. Zappey, was seated on the prow of the boat, and with his field-glasses scanning the cliffs from time to time. " This looks ideal country for Goral," he said to me, standing near him ; " has anyone ever seen them hereabouts?" "I don't know, but there is no record of anyone having shot one," I replied. Scarcely had the words left my mouth when Zappey quietly said, " There's one ! " He rushed into the cabin and secured his rifle ; meanwhile the crew shortened sail. The animal stood under the lee of a cliff some 500 feet above the river ; it was about 4.30 in the afternoon. There was considerable weigh on the boat, and Zappey's first shot struck a little above and in front of the Goral, and the beast scarcely heeded it. The second shot was again a little high, and immediately in front, and the animal swung round, ran a few yards, and then stopped, half facing us. The third shot found its mark ; the soft-nosed bullet passed anglewise through the jugular vein far into the body, and the Goral sank stone dead in his tracks. It was a pretty shot, and,

from the motion of the boat, not an easy one. But the good
fortune did not end here. This Gorge is some 30 miles long,
and throughout its entire length there are scarcely half a dozen
places where it is possible to scale the cliffs to any height
above high-water mark. This was one of the few places!
Willing feet rushed up the cliffside, and in about twenty minutes
the Goral was landed on deck. It proved to be a fine male.
Our crew were delighted, and the incident afforded them
conversation for days. They did not allow the result of my
companion's prowess to remain at one trophy. By the time
we reached Chungking, rumour, having fleet wings, had
reported a bag of five! The fact will probably form the basis
for a legend in these parts, in which at least a score of Goral
will be substituted for this solitary trophy.

The "Ichang Goral" (*Næmorhedus henryanus*), as it
may be styled, has the sides of the body dark greyish; whole
of the tail, front part of upper fore-leg, and line down centre of
back, blackish; fore and hind legs from knee and hock to hoof,
light chestnut; and throat-patch, pale buff. The largest male
shot by Zappey gave the following measurements: length,
46⅕ inches; tail, 4⅜ inches; height at shoulder, 24⅛ inches.

Returning from Tachienlu on 30th September 1908 Mr.
Zappey secured two of the "Grey Thibetan Goral" (*N. griseus*).
Here is his account from a letter he wrote me immediately
afterwards: "At Liuyang, some 10 miles below Tachienlu, I
got two Goral on the cliffs across the torrent with one shot.
The bullet passed through the neck of one and through the body
of the other. I only saw one when I fired; both dropped stone
dead. It was a hard and difficult job retrieving them, taking
us nearly two hours climbing and circumventing cliffs. This
Goral differs from those secured in the Gorges in being much
lighter in colour all over—the legs are light creamy buff instead of
light chestnut; the head is dull grey with a black line from top
of eye to the horns; the throat is very light coloured." The
largest, a female, gave the following measurements:—total
length, 51 1/16 inches; tail, 4⅜ inches; height at shoulder, 25 1/16
inches. The animal is rather larger than the Ichang Goral,
but the tail is the same length.

On the cliffs bordering the Tung River near the base of Wa

THE ICHANG GORAI. ♂, ♀ AND YOUNG ♂

shan Zappey shot, in June 1908, a Goral which he thought looked different from either of the above. Unfortunately it fell on a ledge and could not be retrieved. Possibly this was referable to the Ashy Thibetan Goral (*N. cinereus*), which is distinguished from the foregoing species by its nearly uniform, distinctly ashy colour ; the whitish patches on the throat and feet smaller ; tail longer and more bushy.

Though of quiet colour, Goral are pretty little beasts, and their heads make neat trophies. In a general way they look like small Serow, having similar but smaller horns, and a rather coarse shaggy hair, with a wool-like under-fur, but they have no mane, though the hair along the back of the neck is somewhat crested. They make a curious, rather penetrating, hissing noise when alarmed, and in early April, at any rate, this noise is commonly heard when traversing their haunts. Unlike their near allies, the Serow, they are comparatively social animals, and several are usually found together. The native name is " Yeh Yang-tsze " (Wild Goat) or " Ai (Ngai) Yang-tsze " (Cliff Goat). The flesh is dark coloured and moderately good eating, far superior to that of the Serow or Takin.

The " Grey Goral " ranges up to 8000 feet altitude in the summer, but comes lower down in winter. The haunts are always scrub-clad cliff-country, and it does not appear to frequent timber. The geographical range is considerable, being apparently limited only by the nature of the country up to the altitude given. To my knowledge this animal extends from Lungan Fu in the north to Tachienlu in the west and Wa shan in the south. Goral also occur in western Yunnan, and extend down to Burmah, where the species is different. The probability is that Goral are common to all the precipitous country between 1000 and 8000 feet from western Hupeh, through western Szechuan and southwards to Burmah.

TAKIN, " YEH NIU " (WILD CATTLE)

Few animals have attracted more attention during recent years than this strange and interesting ruminant. The existence of this animal in Western China has been known these many years, but it was not until 1908 that specimens were

authentically shot by a foreigner.   L'Abbé David secured the
first examples of this race through native hunters in 1869, from
the quasi-independent principality of Mupin.[1]   In 1893-4 a
Russian traveller, M. M. Berezovski, secured specimens from
the Kansu-Szechuan border.   I have no precise information as
to just how this traveller obtained them, but I was told, when
travelling through this region, that natives shot them and sold
them to him.   But this is all quite modern.   Marco Polo heard
of these animals during his travels in this region and speaks
of them as " very wild and fierce animals " under the name
of " Beyamini," probably having in mind the wild cattle of
Bohemia.

In the *Field* newspaper of 15th July 1905 appeared an
article under my signature drawing attention to the game
animals of Western China, and the Takin in particular.   This
article attracted attention, and two or three sportsmen visited
that country in quest of this animal.   Ill-health caused one of
them to abandon the enterprise when nearly on the ground.
A second was boycotted by Chinese officials at Tachienlu, and
his expedition rendered abortive thereby.   In 1908 I was again
in Western China, and I invited my friend, Captain Malcolm
M'Neill, to join me and try and secure this animal.   He came,
and success crowned his efforts.

In 1908 there were three distinct parties after this Chinese
Takin, and each party secured trophies.   The first specimen
authentically shot and killed by a foreigner fell to the rifle of
Mr. Walter R. Zappey on 27th May 1908.   Mr. C. H. Mears,
the companion of the ill-fated Mr. J. W. Brooke, followed closely
on Mr. Zappey's heels, securing his trophy on or about 30th
May.[2]   In August, Captain M'Neill, shooting in the petty state
of Yutung with hunters supplied by the Chiala chieftain,
happened upon a herd in open country, and killed several.   In

[1] Much confusion has arisen through wrongly calling this country Eastern
Thibet.   Politically the region belongs to China proper, forming part of
Szechuan province.   Since the boundary generally is ill-defined on all maps,
and the country peopled largely by non-Chinese races, the term Chino-
Thibetan borderland (see *ante*, Vol. I, Chap. XII) may be employed for the
entire region.   But it should be remembered that, so far, every specimen
of the Takin has been taken in China and none in Thibet proper.

[2] See *Life of John Weston Brooke*, by W. N. Ferguson, p. 136.

September, Mr. Zappey, shooting south-west of Tachienlu, secured a fine female, and most unfortunately lost a large bull, which he had knocked over and left for dead.

Since fictitious claims to shooting and many other erroneous statements have been made in regard to the Chinese Takin, I have thought it well to place on record here the names of the sportsmen who shot the first specimens of this interesting animal. Up to Christmas 1910 no other specimens had been shot and retrieved by a foreigner.

This Takin has a wide range in Western China. To my knowledge it is to be found from south of Tachienlu north to the Kansu border, and from this point east into Shensi province, where it occurs on the Tsin-ling range. In certain places, like the wild country between Lungan Fu and Sungpan, the Pan-lan range, and in the petty state of Yutung, it may be said to be common. Anywhere in these regions where there are " salt-licks " this animal is to be found. In western Szechuan its eastern limits are the high ranges forming the western boundary of the Red Basin. As to how far southwards this animal occurs we are without precise information, and it is possible that it traverses the whole mountain-system down to Assam. However, since it has not been reported from Yunnan, it may well be that the southern limit is marked by the Upper Yangtsze, where it makes its great bend to the east.

It frequents difficult country between 8000 and 14,000 feet altitude, making its home in the dense Rhododendron and Bamboo thickets in fairly open forests near the upper limits of the tree-line. It took Mears, as he himself told me in Chengtu city soon afterwards, three weeks' hard hunting in the fiercest and roughest of country to secure his trophy. Mr. Mears said that he shot big game in many lands, but the quest of the Takin proved the hardest and most difficult he had ever experienced. M'Neill, on the other hand, happened on a herd in open country where one would expect to find sheep, but thick cover was not very far off.

A powerful animal, the Takin has no difficulty in forcing its way through dense thickets, tramping out well-defined paths which are regularly used in the passage to and from grazing grounds and salt-licks. Advantage of this habit is

taken by native hunters to spear this animal. Two trees
growing side by side are selected, and a large, heavy log-beam
is attached to a pivot resting in the fork of convenient branches.
This beam measures about 8 feet in length, and in the extremity
a stout stake about 15 inches long and shod with a barbed spear
some 8 inches long is fixed. From the end nearest the pivot
a bamboo rope is suspended. The beam is poised by pulling
down this rope and attaching it to a cunningly arranged contrivance some 14 inches above the ground (see illus. p. 170).
To the stout fixed parts are arranged two collapsible rods, to
one of which a trip-rope is attached. This trip-rope is stretched
across the " run " and lashed to a tree on the opposite side,
the height above the ground being about the same as the
animal's knee-height. The whole trap is a rough and strong
yet a delicate and devilish contrivance. An animal coming
down the run touches the trip-rope with its forelegs, and is
immediately impaled by the spear. The cross-beam, to which
the spear is attached, is so heavy that the spear is driven
almost through the animal's body behind the shoulder, in-
flicting a mortal wound. Death may often be slow, but it is
always sure, and very seldom can a wounded beast break
away. The " run " is only roughly trampled out, and the
bamboo stems and other brush effectually hide the trip-rope.
These traps are in common use, and are a source of considerable danger to anyone traversing these runs. A Chinese youth
employed by Zappey accidentally released the trip-rope of one
of these traps, and the iron spear-head passed right through
the thick of his thigh, luckily missing the arteries and bones.
The spear-head (see figure, p. 170) was cut off on the inside
of the thigh, and the shaft tugged out on the opposite side.
The youth recovered, but suffered a very bad wound for many
weeks.

Dead-falls are also employed by the natives in trapping
many animals, but these are scarcely sufficient to kill a large
adult Takin. These dead-falls are fitted with a treadle arrangement, and the animal stepping upon this causes the whole mass
to fall, crushing him to death.

Around Wa shan the Takin is killed by an arrow shot
from a cross-bow fixed by hunters alongside the run or by

an ingenious gun device. It is also captured by cunningly arranged foot-snares.

The Takin retires during the daytime into thickets, and feeds during the early morning and late evening in the open country above. On misty, foggy days they may feed in the open all day long. During August and early September, at any rate, cows and young bulls are found severally together in the more open country high up, where there is good pasturage in close proximity to thick cover. Old bulls are solitary and wander considerably. The rutting season is around the end of July, and the calves are born in the March following.

The natives regard this animal with considerable awe and dread, affirming it is both fierce and revengeful. Near where Zappey shot his cow a hunter had been killed only a few weeks previously. Doubtless this animal, when wounded, is a nasty customer at close quarters, especially to anyone armed with nothing more effective than a native gun and spear. A foreigner armed with a modern rifle of high velocity, and exercising a moderate amount of caution, has nothing to fear. Accompanied by a native hunter, Zappey shot and killed two calves, and the mother immediately made off through the bamboo jungle, and did not attempt to attack the hunters. These calves were secured in the neighbourhood of Wa shan on 27th May, and according to the natives were about two months old. The legs were soot-black; ridge down centre of back, black; sides, brown; under-parts, black with long whitish hairs interspersed; pelage, woolly. Both were male and, in size, nearly equal, measuring: length, $38\frac{4}{5}$ inches; tail, 3 inches; height at shoulder, $22\frac{1}{2}$ inches; no vestige of horns was discernible.

The adult Takin is a rather awkward-looking, clumsily built animal, strong and powerful, weighing 5 to 6 cwt., or more. None of the animals shot by the sportsmen mentioned above was weighed. The cow killed by Zappey fell in a bad place, and six men could not turn her over, but only roll the carcase from side to side. The skinning under the difficulties took six hours. This animal gave the following measurements : total length, $80\frac{1}{2}$ inches; tail, $6\frac{3}{8}$ inches; heel, $16\frac{1}{4}$ inches; height at shoulder, $42\frac{1}{3}$ inches; height at hip, $39\frac{3}{4}$ inches. The nose, chin, and face half-way up to eyes, area around eyes, tip to tail, hock and legs almost up to knees, black ; ears, greyish ; rest of body light creamy-white ; fore-part of body clearer and lighter than hind-part, which is mixed with greyish hairs. This animal was killed on 17th September 1908, and contained a fœtus about the size of an ordinary squirrel. The adult males are more orange-yellow in colour, particularly on the neck and shoulders, with a dark stripe extending from nape of neck to withers. The bulls run rather larger than the cows. The horns are alike in both sexes, though rather smaller in the female. These horns somewhat resemble those of the Blue Wildebeeste, and are jet-black. They grow downwards and outwards for a short distance, then take a sharp curve upwards and backwards. A pair of horns purchased near Wa shan, and in my possession, measure : length, $20\frac{3}{4}$ inches and 20 inches ; circumference at base, $11\frac{1}{2}$ inches ; tip interval, $11\frac{3}{8}$ inches ; widest spread, $16\frac{3}{8}$ inches. This is the largest pair I have seen in Western China.

Zoölogists attach considerable taxonomic importance to colour, but in the Chinese Takin they must be prepared to grant a wide range of variation. I have seen probably 100 flat skins in addition to the specimens killed by the sportsmen mentioned above. Nearly all showed some distinct coloration, and hardly two were alike, whilst the extreme forms look very dissimilar. The general colour of the bulls may be put down as tawny-grey and black, with shoulders and neck bright golden-brown ; the mane is grey. The cows are much lighter grey, and the older ones are almost white in their upper-parts.

The curved nose, short, almost square ears, and minute stump-like tail give this animal a strange and most distinct appearance. The limbs are very short, thick, and muscular, and the lateral pairs of hoofs are very large. The flesh is dark coloured, and in inferiority of flavour surpassed only by that of the Serow. The animal is hunted for its meat, which is esteemed by the natives, my poor opinion notwithstanding. Flat skins are commonly used in bed-mattresses, and also frequently made into leather. The horns are used as powder-flasks by hunters.

This Chinese Takin was originally described as *Budorcas*

THE CHINESE TAKIN (BUDORCAS TIBETANUS) ♀ SHOT BY WALTER R. ZAPPEY

taxicolor, var. *tibetanus*, by the late Prof. A. Milne-Edwards (*Recherches pour servir à l'Histoire naturelle des Mammifères*, 1874, p. 367, plates lxxiv. and lxxix.). Since then two other names have been applied to this same animal. In the *Proceedings of the Zoölogical Society* (pp. 795–802, with plate), published April 1909, Mr. R. Lydekker considers it a species distinct from the Mishmi Takin, and named it *B. tibetanus*. This seems the most expedient thing to do, but it is unfortunate that the laws of priority necessitate the keeping up of the name " *tibetanus*," which is a misnomer, in preference to the name " *sinensis*," which is both accurate and descriptive from a geographical standpoint. The natives everywhere in Western China designate this animal " Yeh Niu "; this may be translated " Wild Cattle," though " Wild Cow " or " Wild Ox " is an equally correct rendering. Old, solitary bulls are occasionally spoken of as " Ta Yeh Niu," " Large (Great) Wild Ox." Baber (*Royal Geographical Society, Supplementary Papers*, vol. i. p. 39) calls it " Ngai Niu," " Hill (Cliff) Cattle," but has evidently confused it with the Serow, his account unconsciously covering both animals. The people around Wa shan (where Baber collected his information) seem to confuse these animals strangely. In 1904 they insisted to me that the Takin was called " Pan-(or Pngai)-yang " (see article in *Field*, loc. cit.), a name which properly belongs to the Bharal, an animal not found anywhere around that immediate neighbourhood. Experience teaches one to be very cautious in accepting vernacular names, most of them at best are purely local in application, and the natives will readily invent a name to satisfy an inquisitive foreigner.

I have written at length partly on account of the great interest which attaches itself to this remarkable animal and partly in the endeavour to correct certain erroneous statements and misrepresentations which have appeared in reference to the Chinese Takin.

GOA (THIBETAN GAZELLE)

The Thibetan Gazelle (*Gazella picticaudata*) gets as far east as Tachienlu. Indeed, its eastern limit may be put down as the snowclad barrier ranges which, running almost due north and south, are a feature of the Chino-Thibetan borderland. In early summer Goa are found in small herds at elevations of between 14,000 and 17,000 feet, in open moorland country backed by perpetual snows. Later in the season the bucks separate into parties of two to five head. Goa are wary animals, difficult to detect when stationary, since their coloration harmonizes closely with their surroundings. When fired upon they generally run a short distance, then halt, and commence to feed again.

The summer colour of the head and back is grey (in winter light fawn with a grizzly tinge) ; under-parts, white ; on the buttocks the white area forms a large, conspicuous patch ; tip of tail, black. The male has black horns, nearly erect for a short distance, then curving sharply backwards, the extreme points deflected upwards, with transverse ridges closely crowded together. Horns 13 inches long are good specimens, and above the average. These horns are commonly used by muleteers to fasten the bit through in bridles for ponies and mules. A full-grown Goa stands about 24 inches high at shoulder ; the flesh is said to be good eating.

CHIRU (THIBETAN ANTELOPE)

This animal (*Pantholops hodgsoni*) scarcely enters the region with which we are concerned. It is said to occur north of Litang and in the Thibetan province of Derge, on the alpine regions bordering the limits of vegetation. The horns of the male are extremely handsome, being erect, slightly curved, sub-lyrate, jet-black, 20 to 26 inches long, with very fine grain and a number of bold, transverse ridges in front, smooth behind. These horns are occasionally to be seen on sale in Tachienlu. They are used to form the resting-fork on guns of first-class workmanship, and every Thibetan of wealth and property possesses one or more such guns.

It is generally assumed (and with very good reason) that this animal gave rise to the legend of the Unicorn. Among the tribesfolk inhabiting the little-known region south of Tachienlu down to Yunnan, the belief in the existence of the

Unicorn is general ; they declare that it frequents the country immediately to the west. A friend of mine who accompanied the Younghusband Expedition to Lhassa saw a number of Chiru, and he assured me that when seen in profile nearly every animal appeared to have but a solitary horn. The Unicorn figures prominently in Chinese mythology, where, under the name of " Ki-ling," it is placed at the head of all hairy animals. Its influence is always benevolent, and it appears at the birth of men destined to become great sages and wise beings. Its last advent was at the time of Confucius's birth (552 B.C.) !

DEER

Of Cervus proper three species at least are found in the forested regions of the Chino-Thibetan borderland, distinguished by the natives as " black," " white," and " red " respectively. These Deer are sadly persecuted for their horns when in velvet. Fortunately, it is the males only that are so keenly sought after, otherwise they must have become extinct ere this. The full extent of this dreadful trade it is impossible to determine, but the following figures will give some glimmering of its enormity. In his *Report on a Journey to the Eastern Frontier of Thibet* (presented to both Houses of Parliament, August 1905), speaking of the trade of Tachienlu (p. 80), Sir Alexander Hosie says : " Deer horns in velvet, to the value of Tls. 30,000, are exported annually." Earlier (p. 38) he gives the value of these Deer horns as 2 to 20 rupees per catty (1 catty = 1⅓ lb. English ; Rs. 1 = Tls. 0·3 approx.).

In his " Journey to Sungpan " (*Journal China Branch, Royal Asiatic Society*, 1905, vol. xxxvi. p. 38), Mr. W. C. Haines-Watson gives 1500 catties of Deer horns in velvet, valued at Tls. 30,000, as the annual export from Kuan Hsien. Again, on page 41, he puts the annual export of Deer horns in velvet from Sungpan at Tls. 15,000. . There are other places like Chungpa, Kiung Chou, and Sui Fu where a large annual export of Deer horns in velvet obtains, but no figures are obtainable. However, the above is sufficient to indicate how great a slaughter of stags there must be annually in these regions.   At the lowest estimate at least a thousand stags are killed every year for their horns in velvet.

The Chinese consider these horns, called " Lu-jung," an extraordinarily valuable medicine, possessing wonderful tonic and aphrodisiac properties.   This is evidenced by the almost fabulous prices they will pay for them.   In the Imperial Maritime Customs returns for 1910, under Hankow, is the item : " 93 pairs of young Deer Horns ; value, Tls. 8090." (Tl. 1 =2s. 9d. approx.)   Western pharmacologists may say there is no virtue or medicinal value in these horns, but John Chinaman believes otherwise, and is willing to pay the price, high and extortionate as it may be.

The leg sinews of these Deer are also of considerable medicinal value and are exported in quantity from the far west.   Shed horns are valued for making medicinal glue, used in mixing pills, etc.   There is a large trade in these, the annual exports from Tachienlu alone being estimated at 30,000 catties, valued at Tls. 8500.

In every medicine shop of note, in every village and town throughout the length and breadth of China, Deer horns are in evidence.   In Szechuan and other wealthy regions they are abundantly so.   If one inquires in the east and central parts of China where they come from, the answer received is invariably Chungking and Yunnan.   At Chungking it is always Yunnan and Thibet.   West of the Min River one begins to close up to the question pretty quickly.   Coolies laden with Deer horns are frequently met with on all the roads leading from the far west of Szechuan.   Tachienlu, Sungpan, and other towns mentioned on page 161 are all trade entrepôts, and are fed from the surrounding country.

The highlands of Thibet proper probably contribute to this trade, but the headquarters is the wild, almost unknown, region lying between the Upper Min River, the Chiench'ang Valley, and the frontier of eastern Thibet.   This is a region of high mountain ranges where virgin forests of great size still remain.   The upper limits of these forests are the home of these Deer.   These haunts are very difficult of access, and very few foreigners have had opportunity of shooting these Deer, consequently information is most meagre.

The Black Deer, "Hei Lu-tsze," is the Szechuan Sambar (*Cervus unicolor dejeani*).[1] I believe I am correct in stating that this animal is known only from its horns, no skull or skin having yet been received by Western scientists. The horns of this Sambar can be seen in any large medicine shop in Chung-king, Sui Fu, and other cities, and are said to come from Yunnan. But this is only partly correct. This Sambar occurs west of Tachienlu around Litang, and northward at least as far as the high mountains west of Lifan Ting. Unlike the type, this race frequents cold regions, and is in all probability a distinct species. Captain M'Neill saw, west of Tachienlu, a hind and one calf. He describes the hind as looking very black, so much so that in thick scrub for a moment he mistook it for a bear. This Sambar is undoubtedly rare in these regions, but it is remarkable that a race should be found so far north.

WAPITI (RED DEER)

This is the "Hung Lu-tsze" (Red Deer) of the Chinese, the "Ghwar" of the Thibetans around Tachienlu, and is perhaps the commonest of the three species found in these regions. It ranges from the Yunnan border northward to south-western Kansu and possibly beyond. The local chief of Chiala, residing at Tachienlu, keeps several in captivity at his summer palace (so-called) a few miles outside the town. These animals are about the size of a large donkey, and the stags carry fine horns, as witness the illustration (p. 164). The winter coat is light grey, and the summer coat rufous-brown, with a light rump-patch. There is no record of any foreign sportsman having shot this Wapiti, and its identity is uncertain. Possibly it is a local race of the Thian shan Wapiti (*C. songaricus*). Captain M'Neill, when shooting west of Tachienlu, saw a few hinds but no stag, and suggests it may be the Asiatic Wapiti (*C. asiaticus*). In 1904, in open forested country three

[1] Named after Père Dejean, a Roman Catholic priest formerly stationed at Tachienlu. He first arrived in that neighbourhood about 1870, and never left it, dying there in 1906. He trained natives to collect Butterflies, Moths, etc., and he sent to Paris very large collections. In many ways the late Père Dejean was a remarkable man, kindly, courteous, and noble in character. It is most fitting that this fine Sambar commemorate his name.

days west of Tachienlu, I got a fleeting view of two or three of
these Deer, but being taken by surprise I noted nothing beyond
their size and general colour. The antlers here figured I
purchased in Sungpan in 1903. They weighed 11½ catties,
and were fresh from an animal killed a few days previously
only a few miles west of that town. Unfortunately, I lost
them, with other trophies, through a fire in 1909.

WHITE DEER, " PEH LU-TSZE "

The first specimen of this Deer was shot by Captain M'Neill
west of Tachienlu, in 1908, and it proved to be a new race. It
is described and figured under the name of *C. cashmirianus
macneilli*, by Mr. R. Lydekker, in *Proceedings of the Zoölogical
Society of London*, published October 1909. M'Neill kindly
informs me that in height and size this animal approximates
to the American Wapiti. The colour is creamy whitish-grey,
but some are darker than others. He shot two hinds, but
was unable to get a stag. He saw some, however, but none
with horns more than 18 to 20 inches in length.

Mr. Lydekker (loc. cit.) describes this Deer as " allied
to *C. cashmirianus*, but much paler and more profusely
speckled, the general colour being grey-fawn, becoming whitish-
fawn on the throat and limbs, and the speckling as fully marked
on the neck and flanks as on the back. No white on the chin ;
but the whole of the under-parts dirty-white, instead of merely
the abdomen. Dark dorsal line stopping short about the midde
of the back." In the absence of male specimens Lydekker
regards the systematic position of this animal as tentative.
Geographically this new race is very far removed from the most
eastern known haunt of the " Kashmir Stag," and time will
probably prove it to be a distinct species.

When examining, near Mao Chou, some loads of shed antlers,
M'Neill pointed out to me several which he thought were those
of Thorold's Deer. I strongly suspect they belonged to the
race which now bears his name. Under its native name of
Peh Lu-tsze I have heard this animal spoken of as far north
as Sungpan, and very likely it ranges throughout the whole
Chino-Thibetan borderland. The type-specimen of this new

ANTLERS OF THIBETAN WAPITI: HORNS OF TAKIN AND SEROW

Deer can be seen in the Natural History Museum, South
Kensington, S.W.

" CRYING MUNTJAC "

This animal (*Muntiacus lacrymans*), known locally as
" Hung Chee-tsze," or simply " Chee-tsze," derives its specific
name from the presence of a large gland below the eye. This
Muntjac occurs in the immediate neighbourhood of Ichang,
but is more common some distance removed from this town,
ranging from river-level to 5000 feet altitude. In certain
places in Patung and Hsingshan Hsiens it is abundant. Quite
a number have been shot above the village of Nanto, situated
on the left bank of the Yangtsze River at the head of the
Ichang Gorge. This Muntjac frequents brush-clad rocky
places and thin woods of Pine and Oak where a plentiful
undergrowth obtains. Scrub-clad narrow ravines and gullies
are a favourite " lying-up " place during the daytime, and it
at all times prefers steep slopes to the more level country.

One method of hunting this animal is to have men hurl
rocks down the steep scrub-clad slopes, with a " gun " placed
top and bottom walking a few yards ahead of the men. But
the general method employed is to use native dogs, and this
is one of the few things that these dogs are really of any use
for, from a foreigner's point of view. These dogs give tongue
loudly, and hound and bewilder the beast until finally he is
caught by them or shot by the hunter. Stationed at the
bottom of some gulley or point of vantage, the dogs being put
in at the top, the sportsman gets his chance as the Muntjac
attempts to make his escape. But the trouble is that the dogs
are seldom well trained, and knowing no discipline go off at a
tangent anywhere, scaring everything for miles around. A
point to be remembered in shooting a Muntjac under these
circumstances, is to run and pick it up immediately. If the
dogs arrive first, woe betide the trophy ; they eat and mangle
the carcass in double-quick time. At best these dogs are very
exasperating, and as often as not cause needless annoyance and
yield no returns. Occasionally one happens on a pack owned
by a keen and able hunter, and then one's efforts are usually
rewarded with some tangible result.

Muntjac are solitary animals, though several may be found within the same square mile. In their haunts they have well-defined tracts which they usually make for when roused. In running they carry the head and neck low and have a rather ungainly motion. They are not fast, though at a pinch they can get through cover at a good speed, wriggling through and attempting to slink out at the bottom in the least-expected place.

The Crying Muntjac stands 18 to 20 inches at shoulder, and the total length is 38 to 40 inches. The body is reddish-brown; top of tail bright chestnut, under-side white; belly and inside of hind-legs white, front and outside of fore-legs brownish, inside buff; the head and neck is yellowish-brown, with blackish lines down face; the whole pelt is smooth and glossy. The horns of the bucks are 5 to 6 inches long, erect, curving slightly outwards, with the apex sharply curved inwards; in adult males a small basal tine is developed. The antlers are shed annually, though in old animals they are occasionally retained over two seasons. The upper canine teeth in the males are protruded downwards, forming two sharp tusks about 2 inches long. The does are of the same colour as the bucks, slightly smaller, with no tusks or antlers. The young are spotted with white. Muntjac-hunting is quite good sport, and the flesh is most excellent eating. A 12-bore, using large shot (A.A. or B.B.), is the best weapon to use after these and other small Deer, it being safer than a rifle.

Muntjac are found scattered through the hilly country all over the province of Szechuan, and are quite common in regions bordering the western limit of the Red Basin up to 7000 feet elevation. Being almost nocturnal in habit they are seldom seen in the daytime, but on wet, foggy days they are occasionally met with. How far south they range I have no knowledge, but I have seen them to the south-east of Tachienlu and around Fulin.

Mupin is the type-locality for *M. lacrymans*, but our specimens all came from western Hupeh, and the town of Ichang may be put down as roughly marking the eastern limits of this species.

The late F. W. Styan, in Wade's *With Boat and Gun in the Yangtsze Valley*, p. 126, reports the killing of *M. lacrymans* around Kiukiang and also as far east as Ningpo, but regards them as strays rather than residents. This Muntjac is in all probability Sclater's variety (*M. sclateri*), the type-locality for which is Hanchou, a prefecture near Ningpo. This Muntjac is the eastern representative of *M. lacrymans* and is distinguished by its light yellow face and black outer surface of the lower fore-legs. The two species are very closely allied and very possibly be only geographical forms of one species. Sclater's Muntjac has also been reported from Anhui province.

In the Chekiang province, in eastern China, a smaller and paler-coloured Muntjac (*M. reevesi*) occurs. A race of this animal has been described from Ping-hsiang, a coal district on the borders of Hunan and Kiangsi provinces, under the name of *M. reevesi pingshiangicus*, but more material is needed. Indeed, this latter remark is applicable to every animal reported from interior China. Nearly every foreigner, resident or travelling in interior China, carries a gun for pot-hunting purposes, if nothing else, and I am sure each and all would willingly assist science did they but know in what particular they could be useful. From personal experience I know the need for some accurate account of the different game-birds and animals of interior China to assist those who are quite willing to help scientists in these matters to the best of their ability and power. To do something towards supplying this want is the object I have at heart.

Another species, the "Hairy-fronted Muntjac" (*M. crinifrons*), occurs in eastern China, and shows a distinct approach to the Tufted Deer (*Elaphodus*). It is described as larger than the foregoing species, "plum-coloured, distinguished by a crest of long coarse hairs on the crown of the head, almost completely concealing the pedicles of the antlers." [1]

TUFTED DEER (BLACK MUNTJAC)

These animals are placed in a distinct genus (*Elaphodus*), but are closely allied to the Muntjac, from which they may be distinguished " by the pedicles supporting the very minute antlers of the males converging above, and not being con-

[1] R. Lydekker, *Game Animals of India*, p. 264.

tinued as ridges in front of the eyes. There are also marked differences in the form of the skull. These Deer derive their name from the tuft of long hairs crowning the head—a character possessed also by some of the Muntjacs. In the males the upper tusks are very large, and in both sexes the hair is remarkably coarse." [1] Tufted Deer, " Hei Chee-tsze " (Black Muntjac), of the Chinese, are rare animals. Of the three species described from China I have personal knowledge of one only, namely, that found in western Hupeh. This species was first shot by Mr. A. E. Leatham, south of Ichang, in February 1904, and described by R. Lydekker as a new species under the name of *E. ichangensis*.

This animal is sparingly scattered through western Hupeh between 3000 and 8000 feet altitude, frequenting similar country to that favoured by the ordinary Muntjac, but not descending to the river-level. In early April 1907, when hunting ordinary Muntjac, I was fortunate enough to shoot a Tufted Deer. It was driven by dogs along a mountain spur towards me, and I shot it with No. 4's as it tried to descend into a low ravine. The locality was near the hamlet of Putze in Patung Hsien, some three days' journey south-west of Ichang, and very near the spot where the type-specimen was shot by Leatham. Some six weeks later, in the mountains of Hsingshan Hsien, in company with Mr. Zappey, I saw two of these animals together, but they disappeared before we had a chance to shoot. During our subsequent travels in these regions we often heard of this animal, but we saw no others. The animal shot near Putze was a female and carried a young, about the size of a small cat, and black in colour. The adult measured 28¼ inches at shoulder ; total length, 67¼ inches. The body and top of tail was brownish-black, stern and under-tail white, fore-legs brownish-black. Being a female it had neither horns nor tusks.

In Mupin, western Szechuan, *E. cephalophus*, the type of the genus, occurs. This species differs from the foregoing in being rather less uniform in colour, and from measurements recorded is apparently a rather smaller animal. But the species are evidently very closely allied. Around Ningpo, in eastern

[1] British Museum, *Guide to Great Game Animals*, 1907, p. 56.

China, a third species (*E. michianus*) occurs. This is a much lighter-coloured animal than either of the above, being deep brown all over except white belly, white tip to ears, and pale line over the eye.

The ears in all three species are relatively large, broad, and rounded. The lateral hoofs are almost rudimentary in character, and the sabre-like tusks in the upper jaw of the males are not turned outwards as in the Muntjac. They have the same "hunched-up" appearance and gait when running, and do not travel fast. The flesh is very good eating.

MUSK DEER, "CHANG-TSZE"

This pretty little animal (*Moschus sifanicus*) is still fairly common throughout the length and breadth of the Chino-Thibetan borderland, but is everywhere sorely hunted for its musk. This highly valued product is secreted during the rutting season by a skin-gland situated on the genital organ of the male. The whole gland is removed and constitutes the "Musk-pod" (Chinese, Hsiang-p'i) of commerce. This Musk (Shê-hsiang) is by far the most important export passing through the border towns of western Szechuan. Hosie (*ibid.* p. 38) says that some 60,000 pods of musk, worth from 20 to 50 rupees each, according to size and quality, are annually sent through the district of Litang to Tachienlu, where they are trimmed and prepared for the Chinese and foreign market. An ordinary pod in a raw state weighs about an ounce, and with its fringe of skin and hair is about an inch across. Adulteration is commonly practised, but Chinese dealers are experts in detecting this. They have many ingenious tests : If the smell is unsatisfactory, or any doubts exist as to its genuineness, a few grains are extracted from the pod and placed in water. If these remain granular the musk is genuine; if they melt it is false. Another test is to place a few grains on a live piece of charcoal. If they melt and bubble on the red surface the musk is pure; if they at once harden and become cinder it is adulterated. The product exported through Tachienlu is esteemed more valuable than that from Sungpan and other border towns.

Hosie (*ibid.* p. 81) puts the annual exports of musk from Tachienlu at over 24,500 ounces, valued at Tls. 300,000. Watson (*ibid.* p. 38) gives the export of musk through Kuan Hsien as 16,000 ounces, valued at Tls. 216,000 ; from Sungpan (p. 41) to the value of Tls. 60,000. Through the Imperial Maritime Customs at Chungking between 40,000 and 50,000 ounces of musk pass annually. For the ten years ending 1901, some 483,174 ounces of musk were exported through the Imperial Maritime Customs at Chungking. But these figures represent only a part of the export, since they do not cover what passed through the Native Customs. In addition to this export large quantities are consumed in the wealthy cities west of Chungking. In the last Decennial Report (pub. 1904) the Commissioner of Customs, Chungking, writes : " The destruction of these animals must be enormous and must lead to their extinction if the present slaughter continues." The figures given above amply justify the commissioner's views.

This much-persecuted little animal frequents the upper wooded country between 8000 feet altitude and the tree-limit (11,500 to 14,000 feet, according to climate), where forests composed of Spruce, Silver Fir, and Larch, with a thin undergrowth and plenty of rocks, obtain. It occurs solitary or in pairs, though in a small area several may be found. It is a very agile little beast, and a favourite retiring-place during the daytime is the upper part of some half-fallen, sloping tree-trunk. Such trees it ascends with ease, and hunters closely examine every one of these trees for the marks made by the sharp hoofs of this animal. It lies close to the trunk and is not readily detected. Among rocks in the forests is another favourite haunt. A 12-bore with S.S.G. or A.A. shot is the best weapon for hunting these animals. The natives trap, snare, and more rarely shoot them. A male shot west of Tachienlu by Mr. Zappey measured : total length, 34 inches ; height at shoulder, 21¾ inches ; tail, 1⅝ inches. Legs, grey ; body, dark brown (back, reddish-brown), speckled with greyish and tawny yellow : head, grey ; front of neck, light grey ; belly, yellowish-brown ; ears, dark grey, except outside edge, which is light brownish-yellow ; upper canine teeth sabre-like, 1½ inches long.

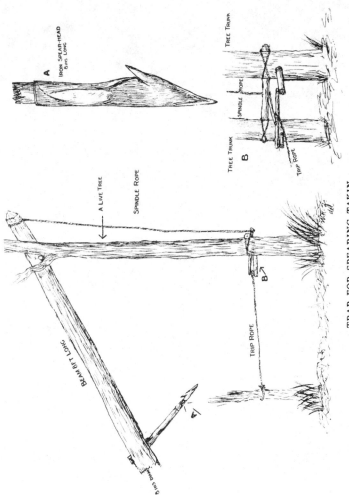

A IRON SPEAR-HEAD 8 INS LONG

TREE TRUNK

TREE TRUNK

SPINDLE ROPE

B

TRIP ROPE

A LIVE TREE

SPINDLE ROPE

BEAM 8 FT LONG

TRIP ROPE

B

8 INS DIAM

A

TRAP FOR SPEARING TAKIN

Neither sex has horns, and the long tusks and musk-gland distinguish the male from the female. The hair is hollow, very coarse and loose, and readily pulls away. The flesh is excellent eating (equal to the best Muntjac), and the heads make pretty trophies. The Chiala chieftain keeps some of these animals in an enclosure at Tachienlu. They appeared, when Zappey and I saw them, to be very happy and contented, and we were informed that they bred under captivity. Certainly they make most charming pets with their shapely face and head.

For a small animal, Musk Deer is of stout and rather heavy build; the hind-limbs are longer than the front ones, raising the rump above the level of the fore-quarters, giving the animal a hunched-up appearance.

The type-locality of *M. sifanicus* is the province of Kansu, and quite likely the Musk Deer found west of Tachienlu represent a local race. Perhaps it should be identified with the Himalayan *M. moschiferus*.

RIVER DEER, " CHEE-TSZE "

It is customary to write disparagingly of this interesting little animal (*Hydrelaphus inermis*), both as to the sport it affords and its value for the table. This attitude may be attributed to " familiarity breeding contempt." For the table there are kinds of venison which are certainly superior, but it is wholesome, palatable, and very much superior to the beef obtainable in most of the riverine ports of China. Properly kept and properly cooked, there are many worse things than a cutlet of this much-abused River Deer.

Formerly this animal was extraordinarily abundant throughout the fluviatile regions of the Lower Yangtsze basin, and it is still very common in many places. It is hunted mercilessly by the Chinese, and several thousands are sold annually in the markets from Shasi down river to Shanghai and elsewhere. The low hills which commence some 30 miles east of Ichang mark the western boundary of this animal as they do of the Ring-neck Pheasant. Its home is the great alluvial plain of the Yangtsze, which extends from the point mentioned above, eastwards 1000 miles to the sea. Any cover is sufficient to hold River Deer, and though it is not averse to water and swamps it prefers the drier land afforded by any rising ground. In winter an ideal spot in which to find this animal is long grass on rising ground near to reed-clad marshes. When the cover is mostly cut (mid-winter) it will be found in open fields lying in the furrows and hollows.

Small shot is usually recommended as sufficient to kill this deer, and so it is at 15 to 20 yards. A charge of No. 8 shot will kill almost any thin-skinned animal a few yards from the muzzle of the gun if it happens to strike a vital spot. A famous big-game shot (the late Mr. H. C. Syers) once killed a Black Panther with a charge of No. 9's when returning at dusk from snipe-shooting. It was a snap-shot at something which crossed the path and entered the brush immediately in front of him. The next day, when he discovered what animal he had shot, he realized the foolishness of his action and the terrible danger it might have involved him in. But this by the way. No danger is to be apprehended from a River Deer, wounded or otherwise, though it is courageous in its own way. I have seen one beat off and wound a Pointer dog almost its own size. There is certainly no sport in killing deer at 15 to 25 yards. Beyond this distance no true sports-man would fire using small shot on the offchance of bringing the animal down. The sportsman is out to kill mercifully and not to maim game.

The only time I have really hunted River Deer was during the winter of 1907–8. Mr. Zappey wanted specimens, and we made a trip down river below Shasi in quest of them. In this flat country a rifle is out of the question, otherwise some excellent sport could be enjoyed. Using B.B. shot we had good sport, bagging every deer but one we fired at. We had men to beat the likely places and to drive the deer across. Most of the bag were killed at about 40 yards, but several fell at over 55 yards and one at 74 yards. Two or three of them shot square through the heart ran 50 to 100 yards before they dropped dead in their tracks.

We limited ourselves to 20, but could have killed many more had we been so minded ; our best day was 9. A couple of heads mounted and in my possession form a pretty trophy, and are a pleasant memento of days spent in the Chinese wilds. Though so abundant, this River Deer is quite rare in museums, and it was the knowledge of this fact that induced us to kill so many.

River Deer stand about 20 to 22 inches at shoulder ; total length, 40 inches ; tail, 3 inches ; heel, 11 inches. The body is tawny grey ; legs and belly, buff ; top of shoulders and rump somewhat chestnut in old males. The hair is coarse and bristle-like and easily pulls out. No horns are developed. In the males the upper canine teeth are protruded downwards, forming scimitar-shaped tusks 2 to 2½ inches long. These tusks are said to develop in old females, but I never met with this phenomenon. The tusks are brittle and easily broken, at least after the animal is dead. The legs are lightly but muscularly built, and the animal can cover the ground at good speed, running great distances and taking to water like a duck. They are prolific breeders, dropping 4 to 6 fawns annually in May. The average weight is 20 to 24 lbs. ; the flesh is dark coloured. Swinhoe, who described this animal, gave it the generic name of *Hydropotes*, signifying " Water-drinker."

This concludes the list of Deer found in the regions coming within the purview of this work as far as is at present known. In the eastern and northern parts of the Empire there are, of course, others. One of these is particularly worthy of mention, namely, Kopsch's Deer (*Cervus kopschi*), found in the province of Anhui, up country from Tatung on the Lower Yangtsze River. Very few foreigners have seen this deer wild, and only one or two have shot it. Captain Malcolm M'Neill made an abortive attempt to secure specimens of this animal, and writes me as follows : " *Cervus kopschi* are very hard to get and extremely shy, being much hunted by the Chinese for their horns when in velvet. They inhabit rough, stony, and brush-clad hills about 4000 feet high, and lie up in the scrub and long grass during the day. Size slightly larger than a Scottish Red Deer, and the mature animals very much darker in colour as a rule. Horns pretty much the same as the Scottish Red Deer—possibly a little longer."

WILD PIG, "YEH CH'U"

These animals are very common through western Szechuan and scarcely less so in western Hupeh, doing great damage to the crops of Irish potato and maize. Repeatedly have peasants almost begged of me to go with them and hunt Pig, but I never felt that a 12-bore and S.S.G.'s (my largest shot) were good enough for a possible encounter with Mr. Pig. When the crops are ripening, the peasants, on the approach of dusk and for several hours afterwards, beat gongs and make all the noise possible in order to scare these animals away. On moonlight nights the din is maintained incessantly the whole night through, and every traveller in Western China must have heard the weird noises emanating from the crop-clad mountain-sides after dark. The natives hunt these animals most assiduously and many are killed annually. The flesh is esteemed and the killing of a Wild Pig is an event for twofold rejoicing.

These animals are nocturnal in their habits, lying up during the daytime in brush and under ledges of rock. They often build large mounds of dry, long grass and sleep under them during the daytime. One day, whilst botanizing over an elevated sloping plateau in Patung Hsien, I chanced on one of these "houses" and was considerably startled by a loud, angry grunt, and just got a glimpse of the black rump of a Pig as it rushed out and plunged out of sight through some bushes. Signs of Wild Pig are everywhere abundant is the mountains, and in any day's march acres of ground can be seen which have been rooted over by these animals searching for succulent starchy rhizomes and roots. Wa shan is a great place for Pig. Hereabouts in 1908 Zappey witnessed the killing of one by three Wild Dogs. He reached the carcass some five minutes afterwards, but it had been disembowelled and rendered decidedly nasty to look upon.

I have repeatedly been told that the flesh of Wild Pig is good eating. I have tried it several times, but consider it decidedly strong and inferior flavoured. Maybe the young are all right. The only young one I ever saw looked reddish-brown from a distance and was too far off to tell if the animal was striped or uniformly coloured.

I assume that the Pig found in Hupeh is *Sus leucomystix*, the species occurring all over eastern China and distinguished by a pale streak on either side of the face. This animal was formerly very abundant in the Lower Yangtsze delta and is found there still in decreasing numbers. Examples have been shot weighing over 400 lbs., but the average weight is between 240 and 300 lbs. In western Hupeh this species or local race, as it very possibly is, ranges up to 9000 feet altitude.

The species found in western Szechuan is *S. moupinensis*, which has short ears and is said to be closely allied to the Wild Pig of Europe and southern Asia (*S. scrofula*). This animal was first secured by l'Abbé David in 1869, from Mupin. It ranges up to 9000 feet altitude and even more, and is abundant from Lungan Fu in the north to Wa shan in the south and Tachienlu in the west. In towns, Mao Chou for example, the flesh is often on sale in the shops. The Chinese are a pork-loving people and esteem the flesh of Wild Pig above that of any other of their wild animals.

HARES

Hares are fairly common in the neighbourhood of cultivation through the whole region coming within our purview, but no Rabbits are found. Around Ichang Hares are fairly plentiful, though they are getting shot out. They keep close to cultivation, and I never met with one in the sparsely populated mountains of western Hupeh above 5000 feet altitude. Around Tachienlu, Sungpan, and other places throughout the Chino-Thibetan borderland a species occurs in situations ranging from 8000 to 13,000 feet elevation. For our purpose, the terms "upland" and "lowland" may be conveniently used to distinguish these Hares.

The upland or "Szechuan" Hare (*Lepus sechuenensis*), as it is called, has the body brown (winter light grey); belly, white; ears, large, brown and grey; a white ring around the eyes. Weight about 6 lbs. A specimen shot by Mr. Zappey measured: total length, 20 inches; tail, $2\frac{3}{4}$ inches; heel, $4\frac{3}{4}$ inches. This Hare looks much bigger than it really is, possibly the long ears give this appearance. The measurements given

above are those of a Hare shot west of Tachienlu ; possibly
that found around Sungpan is a different species.   They look
the same in colour, but my memory of the Sungpan animal
pictures a Hare approximating to the Blue Hare of Europe
(*L. timidus*) in size.   However, I may have been mizzled by
the long ears.

The city wall of Sungpan Ting (alt. 9200 feet) encloses
a mountain-side, the summit of which is 1000 feet above the
level of the town proper.   The mountain-side is largely given
over to terraced fields of wheat, barley, and peas.   In these
fields I have on several different occasions put up these Hares,
and the animal is common throughout the moorlands bordering
the wheat-field area of north-western Szechuan.   It is very
good eating and much superior in flavour to the lowland
species.

The lowland Hares are sadly confused, and it is not
easy to quote names with any degree of certainty.   All those
we collected around Ichang, both north and south of the
river, have been determined as a variety of Swinhoe's Hare
(*L. swinhoei filchneri*).   The colour of the body is tawny above ;
belly, white ; throat, buff ; upper part of tail, black ; lower
part, white.   A large male measured : total length, 20 inches ;
tail, 3¼ inches ; heel, 4½ inches.   The ears are short and the
ends are black, tipped with white ;   average weight, 4½ to 5 lbs.
The measurements show an animal equal in size with the
Szechuan Hare, but it looks very much smaller when running.
In the reed-bed region, where River Deer and Ring-neck
Pheasant occur, the small Chinese Hare (*L. sinensis*) is common.
This animal is about the size of a common English Wild Rabbit,
weighing 3½ to 4 lbs.   The general colour is reddish-brown
with a rufous patch at the base of neck ; ears and upper part
of the tail same colour as back.   This Hare is said to be
restricted to the south bank of the river, but in the region
mentioned above it is equally common on the north bank.

It is possible that other species of Hare occur in these
regions and more especially in central Szechuan (Red Basin).
Sportsmen should closely examine any they may happen to
kill.   The colour and length of tail and ears are good distinguishing characters.

The foregoing are all the grazing game animals yet recorded from central and Western China as far as my knowledge goes. In shops in the town of Wênch'uan Hsien, two days' journey up the Min Valley from Kuan Hsien, and in Tachienlu, I have seen a few horns of a Roe Deer (*Capreolus*), but could obtain no satisfactory evidence as to their origin. However, it is highly probable that some day a race of Roe Deer will be discovered in this little-known Chino-Thibetan borderland. The nature of the country is unsuitable for Yak and other wild-game of the Thibetan plateaux proper.

CHAPTER XIV

SPORT IN WESTERN CHINA

Carnivorous and other Animals, including Monkeys

QUITE a variety of Carnivora occurs in central and Western China, but none is really common, although, in certain places, Leopard and Black Bear have some claims to be considered so. In western Hupeh a few Tiger (Lao Hu) are to be found, and odd skins are brought into the towns for sale at frequent intervals. Nearly every year tales of man- and cattle-eating tigers reach Ichang, and several times foreigners have made futile expeditions after " Mr. Stripes." The rocky, precipitous regions of Changyang and Patung are favourite haunts of this beast. In 1907, when travelling through these districts, I saw some fragmentary remains of clothing belonging to an old woman who had been attacked and killed by one of these animals. A tigress with two cubs had been located in a cave a few miles away, and my companion and self were invited to take part in their death-hunt. To capture the lordly tiger the Chinese collect together to the number of a hundred or more and make a tremendous noise by shouting and beating gongs. When satisfied that the beast is ensconced within a cave they build a large bonfire at the entrance in order to smoke the animal out. All the hunters are armed with guns, spears, knives, clubs, etc., and when the stupefied tiger attempts to escape they make a concerted and bold attack upon him. As often as not he gets away and frequently some of the people get badly mauled. Tigers are also taken in heavily constructed log-traps, partitioned, baited with a live goat, and fitted with a trigger-released door. Another method is by poisoning the " kill."

The Tiger found in Hupeh is a rather small animal, but is

generally broadly and evenly striped, and the fur, though short, is a rich and glossy chestnut-red. The skin of a young male in my possession, which was killed in Changyang Hsien, in the early summer of 1907, measures only 82 inches, total length ; height at shoulder about 22 inches ; yet a more perfectly marked specimen could scarcely be found.

In western Szechuan this animal is very rare, but is occasionally found in the jungle-clad wilderness around Wa shan. In the Chiench'ang Valley and southward through the Yunnan province it is more common and attains to a larger size. From its geographical range this Tiger would appear to constitute a local race which does not ascend to any great altitude, but is thinly scattered through the warmer parts of central and Western China.

Tiger-bones (Hu-ku) are a highly prized Chinese medicine, and are supposed to transmit vitality, strength, and valour to those who partake of them. In the Imperial Maritime Customs Trade Returns of Hankow for 1910 is the following item : " Tiger-bones, 77 piculs ; value, Tls. 6522."

LEOPARD

Two distinct races of Leopard (Lao Pa-tsze) are found in western Hupeh, namely, a lowland variety (*Felis pardus variegata*), distinguished by its darker, more red colour and less bushy tail, which extends inland from the coast to the neighbourhood of Ichang, where it is rare ; and an upland variety, (*Felis pardus fontanieri*) which differs in its smaller size, paler colouring, and more bushy tail. This latter animal ranges westward from Ichang to the Chino-Thibetan borderland, in places being fairly common. The two varieties are found in brush-clad, rocky country, the upland kind ranging to 11,000 feet altitude or more. In western Szechuan, Leopard is scarce north of Mao Chou, but from Mount Omei southward into Yunnan it is prevalent.

This animal is usually taken by the natives in log-traps, as described above for the Tiger ; but occasionally it is noosed in bamboo snares in the same way as Goral. When after the latter animal, near the head of the Ichang Gorge, Mr. Zappey came upon fresh tracks of a Leopard, but lost them again. In the afternoon of the same day he met two hunters who had captured this animal in a bamboo noose and purchased from them the skin and skull.

The Sifan and other tribesfolk are very fond of leopard skins, making them into girdles and robes. On one occasion, when descending the valley of the Upper Min River, I met three men laden with over a hundred of these skins. The men had come from Sui Fu, and were bound to Sungpan. The skins they were carrying came from Yunnan and Kweichou provinces, where the animal is comparatively abundant.

THE SNOW LEOPARD OR OUNCE

The beautiful skin of this animal (*Felis uncia*), known as Hsueh Pao-tsze, can frequently be purchased in Chengtu and Chungking, where it is said to come from Thibet. At Tachienlu and Monkong Ting this skin is more frequently on sale, being brought there from Derge. This prosperous and famous Thibetan state is surrounded on three sides by lofty snow-clad ranges, and these mountains are the haunt of Snow Leopard, which prey on Goa, Chiru, Bharal, and other animals fairly abundant in those regions.

A skin of the Hsueh Pao-tsze in my possession which I purchased at Tachienlu for Tls. 8, measures 94 inches, tip to tip, the tail itself being 44 inches long and extremely bushy. I have seen better-marked skins than this, but never one quite so large. The fur is very soft, long, and thick, of a white ground colour, covered with irregularly shaped black spots, each with a light centre ; on the head the spots are all black. The stomach is pure white and the rings on the part lower of the tail are broad and well defined. Unfortunately, the head was badly skinned and the ears cut away. The Thibetans hunt and trap this animal for its pelt.

The Clouded Leopard (*Felis nebulosa*) is found in Yunnan and Kweichou, and skins are commonly on sale in Chungking and Sui Fu. In size this animal is nearly as large as the Common Chinese Leopard, with a longer tail. The ground colour is pale fulvous grey, heavily covered with large, irregular-shaped, nearly black blotches. The skins are remarkably handsome and rich in appearance.

Lynx skins, locally known as " She-li p'i," are brought in from the Thibetan regions to the north and west, to Sungpan, where they find a ready market among the wealthy Chinese. The pelt is rather dark grey, very thick and soft, and when tanned weighs only a few ounces. They sell in Sungpan for 5 to 7 taels each, according to quality and the supply forthcoming. Possibly this Lynx is a local race of the ordinary Thibetan kind (*Felis lynx isabellina*).

A number of different kinds of Cats occur in Western China, and their skins are commonly on sale in the shops at Chungking, Sui Fu, and Chengtu. The identification of these animals is by no means easy, but the following have been recognized : Chinese Marbled-cat (*Felis scripta*) ; Chinese Jungle-cat (*F. pallida*) ; Fontanier's Cat or Asiatic Ocelot (*F. tristis*) ; Leopard-cat (*F. bengalensis*), and a local race of the Golden or Bay-cat (*F. temmincki mitchelli*). In the mountains of western Hupeh *F. ingrami* occurs. This latter is a rather small tabby-coloured cat ; the head and body measures 19¼ inches ; the tail, 8 inches. It is a particularly vicious animal.

Civet-cats are common all over the warmer parts of central and Western China. Several species occur, but exactly how many is not at present known. The largest and most handsome closely resembles the Indian Civet (*Viverra zibetha*) and may be this species or a local race. It has the same general colouring and alternate white and black rings on the tail, these being usually about nine in number. The other species are smaller and less attractively marked. A dark grey Palm Civet or Toddy-cat (*Paradoxurus* sp.) is also fairly common, and is sometimes kept as a pet.

Residents in Western China interested in natural history would do well to collect skins and skulls of all the smaller Cats, Civets, etc., for our knowledge of the species found in this region is most inadequate. Such a collection would be a boon to systematic zoölogists, and new species and races would undoubtedly be found.

THE PANDA

This richly coloured animal is rare in Szechuan, but more common in Yunnan. In the former province it occurs in the south-west corner beyond the Chiench'ang Valley, frequenting the forested and brush-clad country between 5000 and 10,000 feet altitude. In Chungking, Sui Fu, Chengtu, and other cities the skin is often on sale.

In the shape of its head, short, broad face, and short ears this animal is very catlike ; the claws too are partially retractile. The limbs are short and stout ; the soles of the feet furry ; the tail is 16 to 18 inches long ; stout, cylindrical, and ringed at intervals like a civet-cat. The fur is long, soft, rich, dark, ferruginous on back, shoulders, and flanks ; under-parts, black ; claws, white ; soles of feet, greyish ; forehead, chestnut with rufous stripe running down from the eye to near the snout ; face, lips, edges, and inner surface of ears, white ; outer surface of ears, dark red.

The Chinese Panda ranges from 38 to 44 inches, tip to tip, and weighs 9 to 10 lbs. It is darker and rather larger than the typical Himalayan species, and has been recognized as a distinct race under the name of *Ailurus fulgens styani*. Its colloquial name is " Chu-chieh-liang," which refers to the nine rings on the tail.

THE PARTI-COLOURED BEAR OR GIANT PANDA

This unique animal (*Ailuropus melanoleucus*) is perhaps the most interesting beast found in Western China. Originally discovered by l'Abbé David in Mupin (1869), it was again met with by M. M. Berezovski in the Kansu-Szechuan frontier during 1892–94, but so far there is no record of a foreigner having killed a specimen ; those obtained by the above collectors were taken by natives. Several skins, more or less imperfect, have reached Europe within recent years, but no foreigner has so far seen a living example. The natives of the Chino-Thibetan borderland know this animal well and call it the " Peh Hsiung " (White Bear). In Chinese literature it is referred to as the " Pi." Skins are, on rare occasions, on sale in Chengtu, where they command high prices. In that

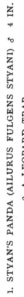

2

1. STYAN'S PANDA (AILURUS FULGENS STYANI) ♂ 4 IN.

2. A LEOPARD TRAP

1

city I have seen in the possession of Europeans several fine examples in use as floor-rugs, but I was never able to secure a specimen myself.

The ears, shoulders, and legs of this animal are black, and black rings surround the eyes ; the rest of the body is rich creamy white. It has a distinct if short tail, and the soles of the feet are hairy. The fur on the pelt is long, glossy, rather soft, and very handsome in appearance. " Parti-coloured " well describes this beast, though from the preponderance of white the native name " White Bear " is very applicable, especially in contradistinction to the Black and Brown Bears of the same region.

The Parti-coloured Bear ranges from the vicinity of Wa shan westwards to the forests beyond Tachienlu, northwards to Sungpan, and thence eastwards through the high mountains to the vicinity of Lungan Fu. It is essentially a denizen of the Bamboo jungles between 6000 and 11,000 feet, feeding on the young shoots of these plants. The natives declare that it eats nothing else, but this assertion is probably too sweeping. Throughout the large area encompassed within the above boundaries, Bamboo jungles are a characteristic feature, forming well-marked zones. In the sparsely timbered belts and in open Silver Fir forests, Bamboo forms absolutely impenetrable thickets. The culms are slender and grow some 10 to 12 feet tall. These plants are impatient of shade from above and grow so thickly together as to starve out all undergrowth and rival shrubs. The young shoots which continue to spring up from June to end of September, according to altitude and species, are white within and excellent eating. The Giant Panda shows good taste in confining his diet mainly to this excellent vegetable !

This animal is not common, and the savage nature of the country it frequents renders the possibility of capture remote. It is occasionally shot by native hunters when after Budorcas and Serow, but is not regularly hunted. It is also sometimes captured in dead-fall traps.

According to the natives, the Peh Hsiung hibernates for the six or seven months in hollow trees, dry, rocky hollows, and caves. Both Mr. Zappey and myself saw evident signs of this animal around Wa-wu shan and south-west of Tachienlu and have every reason to believe that the accounts given by natives are substantially correct. It is a solitary animal, and makes beaten tracks through the forest, frequenting the same haunts for long periods, as is evident from the large heaps of its dung which are often met with in the Bamboo jungle. An adult specimen is said to measure 4 to 5 feet and to weigh about 200 lbs. The furry soles probably form a protection against the splintery stumps of the dead Bamboo culms.

In general appearance this animal is distinctly bearlike, but the skull is much broader than in the Bears proper, and in the teeth and general skeleton it approximates more closely to true Panda. It is the sportsman's prize above all others worth working hard for in Western China.

BEARS

Of Bears proper one species only is common in western China. This is a Black Bear which the Chinese term " Gho Hsiung " (Dog Bear). This animal gets east as far as north-western Hupeh, where, however, it is rare. In the forested regions of the Chino-Thibetan borderland it is fairly abundant, ranging as far north at least as Sungpan. Between this town and Lungan Fu, several days' journey to the eastward, it is prevalent. Around Wa shan in the south and westwards to Litang it is also common. Its altitudinal range depends largely upon the agricultural possibilities of the country, for although this Bear is essentially a lover of rocky forested regions it keeps in the vicinity of cultivation. It is fond of maize cobs and is often surprised and captured whilst feeding on this crop. The limit of maize-cultivation varies from 7500 to 9000 feet, according to climatic conditions. The Dog Bear ranges 1000 feet or so above these limits, and descends to 5000 feet altitude, and even lower in sparsely peopled districts. Starchy roots and the various fruits of the forests constitute the principal food. It hibernates during the winter in dry, rocky caves and in hollow tree-trunks. Cubs are frequently to be seen in June— pretty little black fuzzy fellows, having all the playfulness characteristic of the family. They become ugly with age, however, and treacherous even towards the hand that feeds and pets them. Herr Weiss, German Consul at Chengtu, had a couple which he kept in the Consulate garden for nearly three years, finally presenting them to the Chinese authorities for transmission to the Zoo in Peking. I saw these animals on many occasions and it was amusing to watch them enjoying a bath and frolicking together. Their presence was known to every one in Chengtu, and the local Chinese were much afraid of them and gave their quarters a wide berth. But hunters do not fear them, and the beast is often surrounded by a group of men and killed at close quarters. An adult averages about 6 feet in length, and weighs about 250 lbs. when fat. The fur is jet-black, with a clear white V-shaped mark on the chest and a white spot on the lower jaw ; the muzzle is dark brown, claws dark horn colour ; the hair is long and soft. A specimen secured in western Hupeh by our expedition measured 73 inches from nose to tip of tail ; 40 inches across skin at widest part ; height at shoulder, 38 inches ; hind-paw, 9 inches long, 3½ inches wide ; fore-paw, 7⅝ inches long, 4⅜ inches wide ; claws, 3½ inches long.

Some confusion exists as to the specific identity of this Bear. Captain Malcolm M'Neill shot two specimens near Tachienlu in 1908. A skin and skull of one of these was submitted to Mr. R. Lydekker for determination. In the *Proceedings of the Zoölogical Society of London*, pp. 607–10, with figures (pub. October 1909), Mr. Lydekker discusses this animal and considers it a distinct race of the Himalayan Black Bear,[1] naming it in honour of its discoverer, *Ursus*

---

[1] The oldest name for the Himalayan Black Bear is *Ursus thibetanus*, F. Cuvier, and the reason given by Dr. Blandford (p. 198, *Fauna of British India* ; Mammalia) in rejecting this name in favour of *Ursus torquatus*, " because the animal is unknown from Thibet," is scarcely adequate. It is perfectly true that no specimen, not excepting M'Neill's, Berezovski's, and Mitchell's, has been reported from Thibet proper, and consequently the name " thibetanus " is a misnomer. But if this argument was generally accepted in scientific literature it would be necessary to change a very large number of specific names. For example, the names of *Crossoptilun tibetanum* and *Budorcas tibetanus* would have to be altered, since neither occurs in Thibet proper. Following the laws of priority, therefore, the Szechuan Black Bear becomes *Ursus thibetanus macneilli*. (See Allen in *Mem. Mus. Comp. Zoöl. Harvard*, 1912, xl. No. 4, p. 239.)

torquatus macneilli. He describes it as differing from the type
in the " greater length and softness of the hair, much smaller
size of the cheek-teeth, rather broader skull, and distinctly
vaulted palate which is nearly flat in the typical Himalayan
race." The Szechuan race is founded on a male, and the skull
is in the British Museum, Natural History Branch. This
museum, Lydekker states, " is in possession of a skull of a
female collected by Berezovski," which he also refers to his
new race.[1]

All this is clear and would be amply sufficient to settle the
identity of this animal, but for the fact of the same author
having, at an earlier date, referred a Black Bear from Szechuan
to the Malay Bruang, as a distinct race under the name of
Ursus malayanus wardi, in compliment to Rowland Ward,
from whom he received the skull. This latter (communicated,
I believe, by Mr. Mason Mitchell, erstwhile American Consul
at Chungking) came " from the Thibetan area," and " belonged
to a fully adult Bear of the *Ursus malayanus* type, as is evident
from its width and relative shortness " (Lydekker, *Game
Animals of India,* p. 388).

In 1905 Rowland Ward had received a skull and skin of
a Bear, " reputed to come from either eastern Thibet or the
north-western provinces of China." The skull " was clearly
that of a Bruang," but the skin had " much larger black hair
than the ordinary Malay Bear, with long fringes to the ears,
and the usual whitish gorget on the throat." " The entire
specimen was mounted and sold to the Bergen Museum as
Ursus torquatus " (Lydekker, loc. cit.). And on p. 389 : " The
skin of the Bergen specimen is stated to be more like that
of a Himalayan Black Bear than a Malay Bruang."

From the above it would appear that there are two races
of Black Bear in Szechuan belonging to distinct species.
Without in the least doubting Mr. Lydekker's correct identity
of the skulls in question, there are good grounds for being
sceptical about the occurrence of these two kinds of bears in

---

[1] M. M. Berezovski made his greatest and last collection of animals in the
region of the Kansu-Szechuan border, a little to the north-west of Lungan Fu,
from 1892–94, and in all probability this Bear was obtained there, since it is
common in that neighbourhood.

Szechuan or in the nondescript regions bordering this province and Thibet proper. I have seen in these regions several Dog Bears in the flesh, and a great number of skins, and all were uniform in appearance. Every person I have met in Western China, having personal knowledge of this bear, considered it as the Himalayan Black Bear. Captain M'Neill, who knows the Himalayan animal well, considered the two bears he shot near Tachienlu as identical with the Himalayan race. Mr. Lydekker's examination of M'Neill's specimen confirms this statement in so far as the specific affinity is concerned.

The skull of the specimen secured in western Hupeh and referred to above is in the Museum of Comparative Zoölogy at Harvard College, and agrees exactly with Lydekker's figures of the race *macneilli*. Mr. A. E. Pratt (*Snows of Tibet*, p. 233) says that he secured two bear cubs near Tachienlu : " One reached England alive and was sent to the Zoo." It would be interesting to learn what became of this animal.

The Gho Hsiung is well known locally to natives and foreigners alike as the Common Black Bear of Western China, and it seems scarcely possible that two kinds could be confused under one vernacular name. In the medicine and skin shops of Sui Fu, Chungking, and other large cities in the west, all sorts of curious specimens brought from long distances, and notably from Yunnan, are on sale. There is no reason why the Malay Bruang, or a local race of this animal, should not occur in the warmer parts of southern and south-western Yunnan, and the skulls (like those of other animals occasionally used by fortune-tellers) find their way into the larger cities of Szechuan, after the manner of other products of those regions. I venture the suggestion that the skulls referred by Mr. Lydekker to a race of the Malay Bruang may possibly have had such an origin, and that the skin of the Bergen specimen is that of the Szechuan race of the Himalayan Black Bear. If the specimens were purchased from Chinese sources such a confusion could very easily arise. Whilst I incline to the belief that the race *macneilli* represents the only Black Bear found in western Szechuan, it is obvious that more information in the shape of authentic skins and skulls is needed before the point can be finally settled.

Another Bear is commonly spoken of by the natives of the Chino-Thibetan borderland under the name of " Ma Hsiung " (Brown Bear). This animal is said to be larger and more savage than the Common Black Bear and to frequent the upper limits of the timber-belt bordering the grasslands of eastern Thibet. A gentleman in Chengtu had a reputed skin of this Ma Hsiung, which he used as a floor-rug. Unfortunately I was prevented from seeing this pelt. It was described to me as " dark chestnut-brown, with hair long and coarse."

The skins of the Black Bear are not much valued but are commonly used by muleteers as rough garments and cover-alls, and by peasants and others as bed-mattresses.

Bears' gall (Hsiung tan) constitutes a medicine in considerable repute among the Chinese.

MISCELLANEOUS ANIMALS

Wolves are not unknown in central China, but are very rare, and the same is true in the west also. On the confines of Thibet they are more numerous, and in the grasslands they are common. Quantities of the pelts are imported into Western China by way of Tachienlu, Monkong Ting, and Sungpan Ting. These skins, known as Lang p'i, sell for Tls. 1·75 to 2·50 each, according to supply and quality. The colour varies very considerably, but all that I have seen were, relatively, very pale grey, with the hair on the back tipped with black. The pelage is long, thick, and woolly below. In size these skins vary greatly ; of the several in my possession the largest measures 70 inches total length. This came from Tachienlu. Two species of Wolf (*Lupus filchneri* and *L. karanorensis*) have recently been described from north-eastern Thibet and may range southwards to the vicinity of Tachienlu.

Around Ichang and other places in that vicinity an animal spoken of as the " Dog-headed Fox " (Gho-tou Hu) occasionally puts in an appearance, and is much dreaded by the people on account of its partiality for carrying off small children and goats. Some have supposed this animal to be a Jackal, but skins of two specimens that had been killed near Ichang and brought to us represented nothing other than two old and mangy wolves.

Wild Dogs (Tsai Gho) haunt parts of western Szechuan and quickly drive or kill out all game animals. One afternoon, in 1908, when after pheasants, I saw eight or ten of these beasts within a mile of the hamlet of Tatienchih, situated at the foot of Wa shan. There were three or four together and very brazen, allowing me to approach within 100 yards of them before they slowly moved off. Wild pigs are common in this neighbourhood, and on one occasion Mr. Zappey saw a pig attacked and partly devoured in a few minutes by three of these Wild Dogs. This animal is rather larger than a Fox and decidedly lanky in appearance. The general colour is rufous-grey, the front part of the lower legs being black. The hair is long, and the animal probably represents a local race. Unfortunately, our expedition failed to secure a specimen.

Foxes are more or less common all over central and Western China, and enormous numbers of skins are imported into China from Thibet. The ordinary kind met with in central China has rather short fur, reddish-chestnut in colour, and is of fair size. The Thibetan animal is more fulvous in colour and has longer fur. The skins known as Hu-li p'i are much esteemed by the Chinese for lining silk garments, and are worth Tls. 1·40 to 1·75, according to quality. In Hupeh this animal is colloquially known as " Mao Gho," and this particular species may be *Vulpes lineiventer*, which is common in eastern China.

Three or four species of Fox (*Vulpes filchneri*, *V. ladacensis*, *V. aurantioluteus*, *V. waddelli*) have been reported amongst the trade-skins which enter China from Thibet.

Raccoon Dogs (Gho Wan-tsze) are rather common at low altitudes in central and Western China. Their burrows are frequently met with, and they levy a severe toll on pheasants in particular. On one occasion in Hupeh I killed and marked down a hen pheasant, and on going to retrieve it a few minutes later, found that the bird had fallen opposite the mouth of one of these burrows and had been drawn inside by a Raccoon Dog. The animal found in Hupeh is about the size of a large tom-cat, dark brown in colour, with a short, bushy, ringed tail. I do not know what particular species this kind is referable to, possibly it is *Nyctereutes procyonides*. In western Szechuan a species, pale buff in colour, with the centre of the back and upper part of the tail black, occurs. This has been named *Nyctereutes stegmanni*.

Badgers are not uncommon, and at Ichang Otters are used for catching Fish.

It is not easy to draw the line between game and verminous animals, but several of the above obviously belong to the latter class.

Above the tree-limit in open grassland areas the Himalayan Marmot, or " Hsueh Chu-tsze " (Snow-pig) (*Marmota himalayanus*), is abundant, especially around Sungpan and west of Tachienlu between 10,000 and 15,000 feet altitude. This animal lives in colonies and has a habit of standing on its hind legs at the mouth of its burrow and uttering a shrill noise, half squeal, half whistle. It can be easily shot, but is with difficulty retrieved, for unless killed outright it disappears into the inner recesses of its warren. The male animal is ochre-grey in colour ; the female nearly cream-buff. Adults measure about 28 to 30 inches in length, including the tail, which latter measures about 5½ inches. The fur is coarse and thick, and the skins are an article of commerce in Sungpan, Monkong, and Tachienlu, from whence they are exported to various parts of Western China. These skins, known as Ma-sha p'i, are valued at about 45 tael cents each.

Many kinds of fur-bearing animals of small size occur in central and Western China. These include the Mouse-hare (*Ochotona*), Bamboo-rat (*Rhizomys*), Martin (*Martes*), Flying Squirrel (*Pteromys*), and many species of Squirrel proper. Of these latter I do not remember seeing one frequenting a tree in central China. All of them were rock-loving squirrels. The Bamboo-rat (*Rhizomys vestitus*) is an interesting animal, about 16 to 18 inches long, grey with a white streak down the chest, and has sharp vicious front teeth and powerful jaws. It is common in the jungles up to 8000 feet altitude from Sungpan southwards to Wa shan.

Travellers through the Ichang gorges are commonly regaled with stories of Monkeys roaming in troupes over the cliffs and occasionally throwing stones at passing boats. The stone-throwing proclivities of these animals were once solemnly advanced by local Chinese officials as a reason why foreign steamships should not be allowed to ply between Ichang and Chungking ! I have been through these gorges many times, but have never seen a Monkey. However, it is probable that they do occur in this region. The Chinese are very fond of Monkeys as pets, and have many curious legends concerning them.

A number of species are known to occur in China, especially in the west, where is found the Snub-nose Monkey, or Chinese Langur (*Rhinopithecus*), of which three species are now known. The members of this curious family all have ridiculously upturned noses, tails of great length, and remarkably long and silky hair. The oldest-described species is *R. roxellanæ*, which is fairly common in the forests of the Chino-Thibetan borderland, from the neighbourhood of the Kansu frontier southwards, but more especially in the Chiarung states of Wassu and Mupin and the region lying between Romi Chango and Tachienlu, where it occurs in troupes in the coniferous forests between 8000 and 12,000 feet altitude. In the males the cheeks, throat, sides of the head and neck are bright rusty red, with light patches over the eyes ; crown of head and nape, rich red-brown ; back, grey ; inner sides of the limbs, under-parts, upper sides of the hands and feet, rich orange or bright golden-red ; tail, grey, tipped creamy white ; the bare parts of the face and nose, blue. The female is rather smaller and lighter-coloured in general, with the forehead uniformly coloured bright orange. The male measures about 30 inches head and body, and has a tail about 28½ inches long. The skin of this animal, especially that of the under-parts with the long golden tresses of hair, is used as a lining for garments and worn by the Chinese as a cure for, and preventive of, rheumatism.

In the upper reaches of the Mekong River a second species (*R. bieti*) occurs. The head and body of the male measures about 33 inches and the tail 28½ inches ; the throat, chest, sides of the rump, and flanks are white, the rest of the body is more or less slaty or bluish-grey ; the bare parts of the face bluish-green. The female is smaller, more greyish on the throat and stomach. The very young animals are pale grey, or almost white. The male of this species has a remarkably fine and thick tail.

Quite recently a third species has been described from north-eastern Kweichou under the name of *R. brelichi*. This is apparently the finest of its family, and is one of the largest Monkeys known apart from the anthropoid Apes. It is only known from a flat skin of a female animal, in which the head and body measures about 29 inches and the tail nearly 39 inches. In colour the back is slaty-grey, with a white patch between the shoulders ; crown suffused with yellowish hairs, having black tips ; ears, white ; front of shoulders and inner sides of forearms, deep yellow ; tail, dark with white tip, and longer than in any other known species. The type-specimen is said to have come from " Van Gin shan, about lat. 29° N., long. 108° E." If this is correct the region is several days' journey to the south-south-east of Chungking. Sportsmen-naturalists sojourning at this port would do well to interest themselves in this remarkable animal, for, since only a flat skin is known, it is obvious that specimens (skulls and skins) are much needed.

It is probable that this species is the famous Hai-tuh of Chinese literature of which the following account is given : " Its nose is turned upward, and the tail very long and forked at the end ; whenever it rains, the animal thrusts the forks into its nose. It goes in herds and lives in friendship ; when one dies the rest accompany it to burial. Its activity is so great that it runs its head against the trees ; its fur is soft and grey and the face black."

Several species of short-tailed Baboons (*Macacus*) are found in Western China. L'Abbé David secured in Mupin specimens of *M. tibetanus*, and this is probably the same Monkey common on the middle slopes of Mount Omei and in the region of Wa shan. In the valley of the Upper Yangtsze, near Batang, *M. vestitus* was discovered by M. Bonvalot and Prince Henri d'Orléans. In the valley of the Yalung River, near Hokou (Na-chu-ka), between 9000 and 11,000 feet altitude, *M. lasiotis* is quite common. A female specimen, secured by my companion Mr. Zappey, measured 20½ inches, the tail being 6¼ inches long. The fur is soft and silky, of an olivaceous tint on the head, brightening to nearly clear pale orange-ochraceous on the hips ; feet and fore-limbs greyish.

Monkeys are of course outside the pale of game animals, and many sportsmen strongly object to shooting them. They are, however, of peculiar interest to all, and since more specimens are badly needed, I have thought it advisable to allude to them here. It is not easy to draw the line between game animals and vermin, but those interested should bear in mind that the animals of central and Western China are far from being well known, and that all are rare in museums. Anything and everything of this nature, therefore, is of interest and value, even trade-skins. Sportsmen-naturalists sojourning or travelling in these regions may confer a boon to the science of systematic zoölogy by collecting skins and skulls of all mammals they meet with.

CHAPTER XV

WESTERN CHINA

MINERALS AND MINERAL WEALTH

MINING has been carried on in China for some thousands of years and, in spite of crude methods and the superstition of Fung Shui against which the industry has had to contend, an enormous quantity of mineral wealth has been won from the earth. During the Han Dynasty (206 B.C. to A.D. 25) coal was used as fuel in certain districts, and taxes were levied on iron and salt. Iron mines have been worked in China from very remote times, and this industry, like that of coal-digging, has always been fairly free from official interference. But mining for other minerals and metals has ever been more or less a Government monopoly, and especially is this true of gold, copper, tin, and salt. Intermittently during past centuries, the industry has been vigorously pursued in restricted localities, yet, paradoxically enough, mining in China is in its infancy, for the science of the subject has not been understood or developed, even though the products of the industry have been fully appreciated and generally applied. During the past quarter of a century there has been a real awakening on the part of the Chinese to the importance of developing the mining industry, and it is safe to say that the next fifty years will see the mineral and metalliferous resources of the country tapped and worked on all sides, and mining will assume an importance colossal in comparison with its position in the past.

The mineral wealth of China has attracted very considerable attention from Occidentals during the last decade or two, and concession-hunters have been busy acquiring mining rights and grants from the Central Government at Peking.

These favours have usually been strongly opposed by the provincial authorities, backed by the local gentry, and have involved all parties in endless trouble and difficulty. With one or two noteworthy exceptions these concessions have never been seriously taken up. Vexatious restrictions, official double-dealing, and the opposition of the local gentry have generally proved too much for the foreign concessionaire, and after a time the rights have been allowed to lapse. Western China being so remotely situated from the coast has naturally received less attention in these matters from foreigners, and I am only familiar with one such concession which has been developed by foreign capital. This is a coal mine, located a few miles to the north of Chungking, with which the late Archibald Little was concerned. This notable pioneer eventually disposed of his rights to a Syndicate, which almost immediately, and through no fault of its own, became involved in difficulties with the local gentry. Ultimately realizing the absolute impossibility of developing the purchase satisfactorily, the Syndicate sold it back to the Chinese, which was exactly what the gentry had determined should come to pass.

In Vol. I Chapter VI mention is made of the mineral wealth of the Red Basin, and it is unnecessary to enter more deeply into the subject in so far as this particular region is concerned. Although coal, iron, and salt abound throughout the Red Basin the province of Szechuan is not otherwise rich in mineral wealth. In the south-west corner of the province is found an extension or, perhaps more correctly, a terminative outcropping of the rich metalliferous strata which is so important a feature of Kweichou and Yunnan provinces to the south. These regions supply Szechuan with copper and other metals. Although quantity may be lacking, unquestionably a considerable variety of metals and minerals do occur in western Szechuan, and, whatever wealth in this direction there may be, it is practically as intact to-day as it was a thousand years ago. Chinese antipathy to mine-development is well known, but it is probable that official rapacity and peculation has had more to do with the non-development of this industry than has mere prejudice to disturbing the fabled dragon which slumbers beneath the earth's crust. In Vol. I Chapter XVIII it is told how official exactions had resulted in the closing down of certain copper-workings ; the same story could be told of silver and gold mining in various places throughout the Chino-Thibetan borderland. In several instances Chinese capital has been invested in mines and work commenced, but after a few months the whole thing has been abandoned for reasons usually connected with peculation official and otherwise.

The districts supposed to be rich in precious metals are zealously guarded by official eyes, and foreigners are less welcome guests in such places than elsewhere in China. My business had nothing to do with mines or mining, and it was politic to keep away from any gold or silver workings. I never exhibited any curiosity in this direction, although both in Yunnan and Szechuan I have passed within close range of several mines reported rich in one or other of these metals. But without being obtrusive it is possible to gather information of all sorts, yet this short chapter is written with great reluctance, and only because of its necessity in order to complete the account of these little-known regions.

In earlier chapters mention is made of the rude placer-mining carried on by the unemployed peasantry on the foreshores of all the larger streams throughout Szechuan. The returns are most insignificant, and the industry would never be attempted in lands less overpopulated than China. In the district of An Hsien and in the prefecture of Lungan Fu, both situated in the north-west of the province, gold-bearing quartz occurs and is worked and crushed, but the industry is only on a small scale. In the district of Mienning Hsien, in the Chiench'ang Valley, there is a Government gold mine fitted out with foreign machinery. This Moha mine, as it is named, a few years ago received a development grant of 100,000 taels from the provincial treasury, and intermittent attempts at working it have been made without, however, any substantial returns.

Most of the gold used in, and exported from, Szechuan comes from the western limits of the Chino-Thibetan borderland and from Thibet proper. The district of Litang is one of the principal sources of supply, and there, as elsewhere, it is obtained by placer-mining. In the Chiarung state of Badi-

THE SALT WELLS AT KUICHOU FU

Bawang there is much gold, but it is jealously guarded from Chinese hands. This small State comprises a narrow strip of country on both sides of the upper reaches of the Tung River, thereabouts known as the Tachin Ho (Great Gold River). From time immemorial this and the surrounding regions have been famous for their precious metals.

The aggregate of gold won from these various places must be very great. Practically all has been obtained by placer or pocket mining, and the metalliferous lodes appear never to have been found *in situ*. Possibly these exist in regions more remote and nearer the sources of the Tachin, Yalung, and Dre Rivers. Certain it is that the nearer the head-waters are approached the richer in gold become the sands and shingle-beds of these great streams. The gold is melted and made into small bars weighing 10 or 12 Chinese ounces. A surplus is exported down river either in the form of bars or converted into gold-leaf. The trade in gold is in the hands of Shensi men, as is also the trade in silver. These Shensi men are also largely concerned in the cash shops, native banks, and banking.

Silver-ores occur throughout the Chino-Thibetan border-land from Sungpan in the north to the borders of Yunnan in the south, but though mines are, or have been, worked in many places the region is really poor in silver. The provincial mint at Chengtu is mainly supplied by imports from down river. A certain amount of silver also enters the province from Kweichou.

Copper-ores are much more abundant than silver, and occur from Pêng Hsien southward to Yunnan. In the south-west corner of the province, especially in the department of Huili Chou, copper mining and smelting are considerable industries. The mines are worked by private companies holding licences which compel them to sell the metal at a fixed price to persons duly authorized by the Government. Both the Provincial and Imperial Governments appear to have a controlling hand in this industry, and the companies frequently complain that they cannot work the mines with anything like commensurate profit. Enormous quantities of copper are used in Szechuan for a variety of purposes apart from that of minting cash

pieces of various denominations.   Tribute-copper from Yunnan
is sent down river by way of Chungking to Ichang in native
craft, and at the latter port is placed on board steamers for its
intended destination.

The department of Huili Chou also yields white copper
(Peh-tung) in considerable quantities.   This alloy is produced
directly from the ores, in which the composing metals appear
to be always present in little varying proportions.   The
coppersmiths receive it in round cakes about 7 inches in
diameter.   They remelt and alloy it with copper, zinc, tin,
and lead in varying proportions to suit different purposes.
White copper is used for an infinity of purposes, of which the
making of water-pipes is one of the most extensive.

Spelter or zinc (Peh-yuen) is found from the prefecture of
Yachou southward to the boundaries of the province.   Lead-
ores occur from Sungpan Ting southward to Yunnan, but are
perhaps more abundant in the district of Chingchi Hsien than
elsewhere.   Sulphur is found in the district of Kuangyuan
Hsien in the north, the department of Mao Chou in the west,
and the prefecture of Nanch'uan in the south.   This latter
region supplies the greater part of the province.   The industry
is carried on under Government surveillance, and the retail
price is regulated by the provincial authorities.   Iron-ores
are found all over the mountainous regions of the west, but
except in the prefecture of Yachou are less worked than in
the region of the Red Basin.

Coal is scarcely found west of the limits of the Red Basin,
and antimony, which has recently figured as a large export
from the Hunan province,[1] has so far only been recorded from
one district in Szechuan, and that in the extreme south-east
corner of the province.   A very inferior kind of jade occurs
in the district of Wênch'uan Hsien, where it is mined and
exported to Kuan Hsien and Chengtu Fu, at which places it

---

[1] This antimony is exported through Hankow.   The Imperial Maritime
Customs Returns of that port for 1910 give the following figures relative to this
trade :—

		Piculs		Taels
Crude antimony .	.	157,486	value	900,377
Ore	.	21,909	,,	58,004
Refuse	.	41,568	,,	13,871

is made into bracelets, rings, and other ornaments, and sold at a very low price.

Asbestos in small quantities is found in the prefecture of Kuichou, and also in the Chiench'ang Valley, but the supply is apparently insignificant. Tin has not yet been discovered in Szechuan, and I never heard of any precious stones being found there either. However, the country bounding the Liu-sha River from a little south of Chingchi Hsien to Fulin certainly looks as if it might contain the latter.

From the foregoing brief and scrappy résumé it will be evident to those interested that before any accurate and comprehensive account of the mineral and metalliferous wealth of this region can be written a mineralogical survey must be undertaken by competent persons. It may be that the wild mountain fastnesses of the west contain mineral wealth of great value, yet the probability is to the contrary. That the south-western corner of the province is rich in metalliferous lodes is as far as our present knowledge goes. All the more accessible and populous parts of Szechuan are well supplied with coal, iron, and lime of good quality. The Red Basin is extraordinarily rich in brine deposits, and salt, which is a Government monopoly, is the only mineral at present exported from Szechuan.

The most fascinating subject connected with the mineral deposits of Szechuan is that of the famous " Fire-wells " of Tzu-liu-ching, the gas from which is employed locally in salt evaporation. There is good reason to believe that this gas emanates from petroleum beds which have not as yet been reached. The late Baron Richthofen estimated the coal-bearing ground in Szechuan to exceed in size the total area of every other province of China, though, at the same time, he pointed out that the bulk was too deeply buried to be ever of practical value. Possibly the untapped mineral-oil deposits of Tzu-liu-ching will some day become available and exceed in value that of all other mineral deposits found in the vast province of Szechuan.

CHAPTER XVI

CONCLUSION

SOME GENERAL REMARKS ON THE REBELLION, THE CAUSES
  WHICH HAVE PRODUCED IT; THE PEOPLE AND FUTURE
  POSSIBILITIES

DURING the past two years events in China have
followed so rapidly on one another that it is scarcely
wise to attempt to discuss any phase of the situation
other than that presented at the moment. Much history has
been made in a very short period, and he would be a bold man
who ventured to predict, other than in a general way, the possibilities or probabilities of the near future. Such conflicting
accounts of the progress of events have been published that it
is next to impossible for the man in the street to obtain anything like an accurate idea of the situation. According to a
certain section of the foreign Press the Chinese in dethroning
the Manchu Dynasty emancipated themselves from a verit-
able Kingdom of Belial, and in its stead have established a
promising Utopia. The more reliable section of this Press,
however, takes a rather different view. That there has been
a tremendous upheaval is perfectly obvious, but what the
permanent result will be is undeniably obscure. The progress
of the revolution has been amazingly rapid, and the results to
date have been achieved with a minimum loss of life. Much
privation and suffering has been wrought among all classes,
but on the whole the dynastic dethronement has been more
peaceably brought about than changes of a similar character
in the past.

It cannot be said that the revolution came as an unexpected
event to those intimately acquainted with China. On the
contrary, the surprise was that it had been so long deferred.

Many students of things Chinese fully expected that an up-heaval would follow immediately upon the death of their Majesties, the late Empress Dowager and Emperor Kwang Hsü. During the period of nearly three years of grace which followed their demise, the Central Government became more and more inept, puerile, and rotten; the provincial gentry and the student class more openly rebellious. The sanctioning of a foreign loan by the Central Government, which, among other purposes, provided for the construction of the Hankow-Szechuan Railway and the employment of foreign engineers, was merely the last straw. It is difficult to see how anything could have saved the late Dynasty short of a complete renova-tion of its system of government and the installation of a totally new class of officials capable of honestly conducting the necessary reforms. Such a change was an impossibility, but the Dynasty's life was prolonged a few months by a succes-sion of fair promises which, if made in good faith, it had not the strength to carry into effect.

The late Manchu Dynasty has been equal to any of the long line of Dynasties that have ruled China. But it outlived itself and became an effete anachronism. In accordance with natural law it had to disappear and make way for another more in accord with the times. This Dynasty reached its zenith during the reign of the Emperor Kien-lung (A.D. 1736–96), and on his death it commenced to wane and wane rapidly. But for foreign intervention it would probably have dis-appeared during the great Tai-ping rebellion (A.D. 1850–64). With the merits or demerits of this foreign intervention we are not concerned. The fact remains that the time was only postponed. It gave the Dynasty a lease of life, but all to no purpose, as the recent disasters amply prove. There is, after all, nothing fundamentally new in the present unfortunate condition of affairs. During her long history China has been through it all before, and many times. But the forces which have induced the present rebellion are novel, since in her past rebellions she has not felt the power of Western civilization in the sense that she is feeling it now. The oldest of existing nations, China is attempting to attune herself anew in order to maintain her position as a nation, and to successfully compete with the modern and more aggressive civilization of the West. This latter, which came faintly knocking at her door a hundred odd years ago, has proved an irresistible power, and in spite of the, metaphorically speaking, heavily bolted and barred gates has made entrance, and is to-day well within the citadel. By riots, massacres, and several wars, China has tried her utmost to thrust back this forceful alien, but all to no purpose. The Boxer Rising in 1900 was the final effort. This failed utterly and miserably, and China gave up the contest.

Long previous to this last effort, the most upright and far-seeing of China's statesmen had realized the hopelessness of the struggle, and had begun to urge upon the nation the importance of learning from the Occident all that was useful and helpful in order to renovate their country's condition, and render China strong and able to withstand foreign aggression. The progress and enlightenment on Western lines since 1900 has been nothing short of marvellous. But, unfortunately, the ultra-progressives wanted to run before they could walk, and the ultra-conservatives were scarcely willing to move at all. For the time being the ultra-progressives are foremost, but this can only be a transient phase. Those adhering to the broad, happy medium of holding fast to all that is good of the old and building on it the best of the new must come into their own eventually. The earlier this happens the shorter will be the bitter period of travail. China's essential need to-day is what it has been for a century—a strong central government. Until this is vouchsafed there can be no lasting peace within her borders.

Parliaments have been spoken of as a panacea for all evils of government. If one looks around, very pertinent questions concerning the universal fitness of this system present themselves. In discussing the East it should ever be remembered that the Oriental mind is far from being in complete accord with the Occidental mind. Parliaments are of the West, and the Western model will have to be very considerably altered and modified before it can be successfully employed in the East. A republic, theoretically the highest and best form of government, has not altogether proved to be so in practice, judging by world-wide examples of to-day. When every republican

MY CHINESE COLLECTORS; MEN OF THE PEASANT CLASS

citizen realizes fully the enormous responsibility resting upon him, and acts accordingly, the theory will become accomplished fact. The peoples of the most advanced Western nations are scarcely yet equal to this ; how then can such a form of government succeed in the Far East ? A strong, just man is appreciated the world over. In the East he ranks as a demi-god, and his authority quickly becomes undisputed. China's salvation will not yet be found in any advanced Western system of government, but in a wise, liberal despotism. Granted this, peace would speedily spread throughout the length and breadth of the empire, bringing with it prosperity and content to the industrious, patient, peace-loving millions.

But lest the noise of the revolution, the effeteness of the late Dynasty, and the question of the stability of the present system of government obscure the real China, it may be well to pause for a moment to consider the country itself and the people inhabiting it. The eighteen provinces which make up China proper have a total area of, roughly, 1,500,000 square miles, and form a nearly square tract of country some 23° long. by 20° lat. The size is about fifteen times that of the United Kingdom, or seven times the size of France, or nearly half as large as the continent of Europe. Compared with the United States of America, it is equal in area to all the region east of the Mississippi River with Texas, Arkansas, Missouri, and Iowa added. It is broken up into mountain, valley, and vast alluvial plains, and is drained by a magnificent network of navigable streams. The climate is continental in character and temperate over the greater part of the country. The south is within the tropics, but in the north the winters are almost arctic in severity. Three-fourths of the entire area is well adapted to agriculture, for which purpose it compares advantageously with any similar region in the world. The potential wealth represented in its mineral and metalliferous deposits is beyond computation. Such is the country of China proper without reference to the wealth of Manchuria or the vast area of the Outer Dominions.

The Chinese are a homogeneous race, estimated at not less than 400,000,000 of people, and are without caste pre-judices. They have unquestionably great brain-power, and possess many solid virtues as well as peculiar national defects. They are also an extraordinarily virile and fertile people. So virile are they that as a nation they have absorbed their conquerors as readily as they have done the nations which they have conquered. It is astounding the influence they wield in this direction. As an example, take the late Manchu Dynasty. So completely have the Chinese absorbed the Manchus that to-day they are more "Chinese" than the Chinese themselves. Yet no other nation can absorb the Chinese, and though they are found the world over their nationality is everywhere unmistakable. This virility is likewise exemplified in many other ways. For example, they are apparently indifferent to climate, and are to be found as workers from the arctic regions to the tropics. The Chinese are everywhere the one coloured race which can work and will work. Wherever they are found they are industrious and capable. They wax wealthy where whites would starve, and no nation in matters of labour can compete with the Chinese on equal terms. They can copy anything and everything, and under foreign supervision have already turned out such complex machines as railway engines and steamships.

Malthusian principles have never been listened to, much less practised in China. Large families are everywhere gloried in, and children abound throughout the land. Since a son is necessary to carry out the rites of ancestor-worship, boys are more generally in favour than girls, yet it is a mistake to think that the latter are despised or ignored. Even if preference be shown to sons, the daughters have a share of the family's affection. Infanticide, save in times of famine and dire distress, is not practised, all the stories written about it notwithstanding. Anent ancestor-worship we have nothing to say, except to remark that the respect and esteem for parents and old age, so characteristic of the Far East, is a wholesome example to the Occident.

A keynote to the Chinese character is pride. They are an intensely proud people, and it must be confessed that their pride is justified. Look on their history, their conquests, their inventions, their arts, and crafts. In the history of the world few nations have equal claims to honour and greatness with the Chinese. They have also grave national faults, and this pride and its concomitant conservatism is largely the cause of their present position. Considering themselves the "whole earth," they have persistently and most superciliously ignored the "outer barbarian," as they termed the rest of the world, until disaster upon disaster has shaken the very foundations of their empire. That the scales are falling from their eyes is evident, but as a nation they have yet to grasp the fact that Western knowledge, even though it be of comparative "mushroom growth," cannot be acquired by the study of a few months, and neither can Western institutions be transplanted bodily and in adult form into China. I have met in China hundreds of students intent on acquiring Western knowledge, but scarcely one who in any sense realized the immensity of the task before him. These students persistently refuse instruction in the elementary branches of this knowledge, and are ever clamouring for their instructors to pass at once to the advanced and honours stages. The national defect, pride, is at the bottom of this attitude, and they have yet to appreciate that they must crawl first, then walk before they can safely run.

The merchant class in China is as honourable as that of any country in the world, and foreign relationship with this body has always been satisfactory and mutually advantageous. The artisan, peasants; and farmers are unsophisticated, and every traveller has a good word for them. They are peace-loving, law-abiding; and very easily governed. It is somewhat otherwise with the gentry, students, and officials, who, as a class, have in the main always been more or less opposed to foreign intercourse, and have been the direct cause of many difficulties. For generations China went in for competitive examinations to supply all official posts, and had, as a result, a body of truly incapable officials. The principles associated with Tammany in the West were rampant in this class. Offices were sought and held for personal profit without any regard for public good. But individuals were less to blame than the system which time and pernicious methods had produced. The salaries attached to the various posts were ridiculously inadequate, and the holders had to peculate in order to exist, if for nothing else. Had the Government provided for its officials a " living wage " it could then have expected and demanded an honest, civil, and military service. Instead of this, offices were commonly sold by the Government for large sums, and the purchasers allowed to farm them to their own advantage.

It has been a rule in China that no important office could be held by an official in his native province. Theoretically, like much else in the Chinese system of government, this was an excellent rule, but in practice it had decided drawbacks. All officials, in consequence, were in the nature of aliens coming as they did from other provinces. At best they were strangers and, according to their strength of character, usually became subservient to, or at variance with, the local gentry. In wealthy provinces like Szechuan and Hunan, for example, these local gentry wield enormous power and are, in fact, the real rulers of the province. Local patriotism and self-interest are combined in this class, which commands a large following. The policy and actions of the late Government at Peking were very often antagonistic to the views of these gentry. Especially was this so in the matter of mining-concessions, open ports, and foreign loans. The Provincial Governments were frequently in a quandary in their efforts to harmonize the diametrically opposed views of both authorities. As the Central Government weakened, the provinces became more and more under the power of the gentry. The climax as the world now knows was reached in the early autumn of 1911, when the local gentry of Chengtu Fu induced open rebellion which, spreading with marvellous rapidity, very soon brought about the dethronement of the Dynasty.

The Provincial and National Assemblies which the late Government was virtually forced to call into existence were mainly composed of gentry. The student class, both of the old and new schools of learning, in the main is made up of the sons of this same body. These students are far from being seekers after knowledge for knowledge's sake. In Chengtu they were several times openly mutinous, setting the Viceroy's authority at naught, and compelling him to grant their desires. In many parts of the country the authorities were almost openly afraid of these headstrong students, and

CHAO-ÊRH-FÊNG, VICEROY OF SZECHUAN, MURDERED BY THE
REVOLUTIONISTS

totally unable to check their follies or curb their rebellious ardour.

It is impossible to estimate how much the famous work called the *Chuen Hioh Pien*,[1] written by the late Chang Chih Tung, when Viceroy of Liang Hu (Hupeh and Hunan provinces), has influenced recent events in China. Since 1900 many of the things advocated in this book have been put into practice, especially the matter of schools for Western learning, army reform, and newspapers. But the moral teachings the Viceroy enunciated so earnestly have been set at naught, and republicanism, which he so emphatically denounced, has been brought into being. China for the Chinese was the patriotic vision of this grand old man of China, but perhaps the fates were kind in removing him from this sphere before he had time to see his vision so passionately taken up as a slogan and pressed so hastily forward.

It was fortunate for China that at least one practical statesman remained to take up the reins of government in the hour of need. Yuan Shih Kai is very human, but he has best right to be acclaimed the saviour of China. If his health and strength remain he may, in a few years, weld the country into a solid nation. If loyally supported by those who seek their country's good he can utilize all that is useful and worthy in the schemes of the more visionary ultra-progressive reformers, and render it of practical value. That his hands will be forced at times is certain, but granted time Yuan Shih Kai will succeed in bringing order out of chaos and place the Government of China on a sound basis. At present his position may be likened to that of a skilled driver who has had the reins of a runaway team thrust into his hands from those of an incompetent person. In time he will get his countrymen down to a common-sense trot, and then all danger will be past. If allowed by the passionate rush of events to exercise the rights of his office Yuan Shih Kai will be able to surround himself with advisers Chinese and Foreign, whose interests will be none other than the welfare of China.

The pressing need of the moment is, of course, funds to carry

---

[1] Translated into English, under the title *Learn*, by the Rev. S. I. Woodbridge.

on the Government and to disband the soldiery. This difficulty involves the vexed question of foreign loans, which has
significance far beyond that current in Occidental newspapers.
It is the reef on which the young republic may wreck its
ship of State. To the narrow view-point of the local gentry
and their following these loans will doubtless continue to
appear unessential, and if their interests, real or supposed, are
affected by the security demanded, their antagonism will be
as bitter as formerly. Another difficult class, which if more
evident under the old régime has not yet had time to disappear
completely, consists of officials who desire to handle loans for
the benefits they personally can derive from them. The
competent body of officials who know and appreciate the
absolute necessity of maintaining the country's credit by
meeting all obligations, and also the importance of developing
the resources of the empire, will most certainly have a difficulty
in disposing of the opposition of one party and the mercenary
desires of the other.

That there are financiers willing to lend money on question-
able security and a gullible public willing to subscribe to such
loans is everyday history. Money is absolutely necessary
to carry on this new republic of China, but it is trite to say
that she must beware how and in what manner it is obtained.
The integrity of China is the one thing above all others which
the Government must maintain, and promiscuous borrowing,
even if it tides over present difficulties, may lead to even
greater danger in the near future. The one foreign-supervised
service which the late Dynasty grudgingly became reconciled
to, namely, the Imperial Maritime Customs, has proved the
strongest security possessed by the Central Government.
Further, the international character of its personnel has been
a tremendous factor in maintaining the integrity of the empire.
The present Government will surely be well advised to ponder
this thoughtfully. Until China has evolved a thoroughly competent civil service the example of her plucky and illustrious
neighbour, Japan, is worthy of deep consideration. However
galling it may be to the pride of the Chinese, it would appear
absolutely essential in the best interests of China herself and the
integrity of the empire that all foreign loans should be obtained

through highly accredited international agencies. And, further, that foreign supervision of these loans, their application and disbursement, should be allowed, consistent with the maintenance of the dignity and prestige of the Chinese Empire.

For many years past the Occident has been urging the Chinese to "wake up!" Towards the end of the last century the clamours in this direction became most vociferous. The present upheaval is China waking up—nothing else. The outcome is fraught with colossal possibilities. China needs roads, railways, rolling-stock, machinery to develop her agriculture, mineral resources, and potential wealth generally. All these and very much more she stands in need of. Will the Occident merely furnish the samples and patterns for the Chinese to imitate, and ultimately supply? Or—and who knows? Granted a staple Government, the Chinese people can accomplish much. I do not believe in a "Yellow Peril" in the nature of a possible military conquest of the West. It would be necessary to fundamentally alter the Chinese character in order to make it militantly aggressive. But in their virility and industry they are unconquerable people, quite the equals of the West in these qualities. If they thoroughly "awaken," what is to prevent them becoming in commerce and industry the great competitors of the white race? Time solves and adjusts all problems, and it will this, the real Yellow Peril, if such a thing be within the realm of possibility.

To the people of the Occident in general all forms of civilization other than their own appear effete and retrograde. This is, perhaps, very natural, but will the Western ideal always dominate and remain the criterion, and must all the other peoples of the earth conform to it or become nonentities in the world's future history? The people of the Middle Kingdom will undoubtedly accept from the West and utilize all the material advantages and mechanical appliances that have resulted from the discovery of steam and electricity and may yet retain their own ideals of life. On the tenets of their older civilization these people can wisely build and maybe they will evolve a state of existence intrinsically higher, more restful, and better suited to their nature and needs than that contained in the Western ideal.

China is a continent rather than a country, and everything is so entirely different from and opposite to Western ideas and practice. Hundreds of books have been written on China and the Chinese, yet little more than the fringe of the subject has been really broached. There is, indeed, no finality, and in any one book it is impossible to do more than itemize an occasional fact or two. Nearly eleven years of my life have been spent wandering up and down the by-ways of interior China. I was there through the Boxer crisis and the Russo-Japanese War, and also through certain local riots and disturbances. My experiences in China, though varied, have on the whole been very pleasant. To speak as we find and courageously is the only just stand to take. With all their peculiarities, conservatism, and faults, the Chinese are a great people. Phœnix-like, China has arisen time and again from the ashes of decadent dynasties, and there is every reason to believe she will accomplish this again. Her peace-loving, industrious millions can never be utterly smothered or nationally effaced. Sooner or later they must come into their own, and side by side with the people of the Occident help forward the destiny of the world.

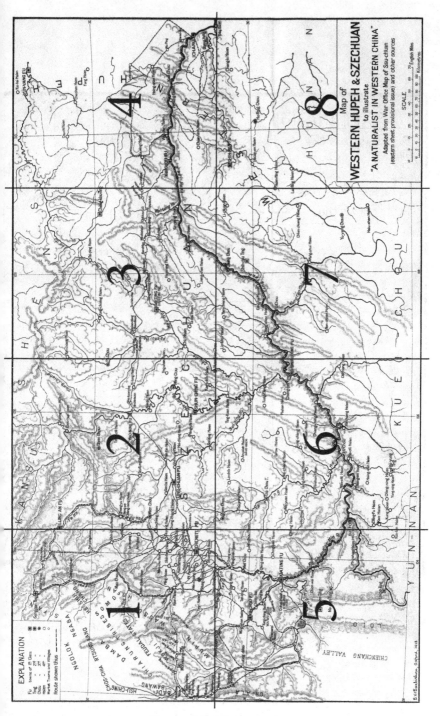

Key to following pages.

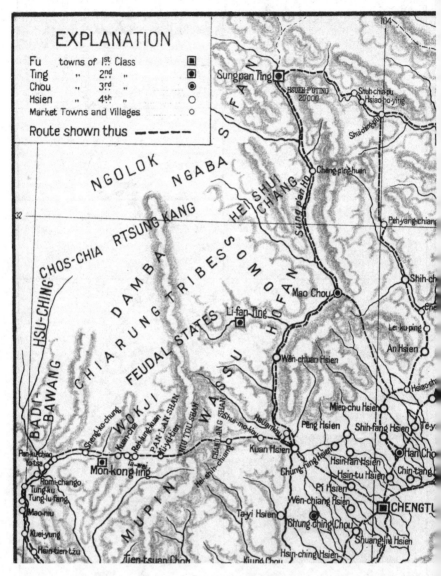

EXPLANATION

Fu	towns of 1st Class	▣
Ting	„ 2nd „	▣
Chou	„ 3rd „	◉
Hsien	„ 4th „	○
Market Towns and Villages		○

Route shown thus – – – – –

104

Sung pan Ting

HSUEH P'O TING 20'000

Shuh-chia-pu
Hsiao-ho-ying

Shui-ching-pu

S I F A N

N G O L O K   N G A B A

Cheng-ping-huan

Peh-yang-chiang

32

R T S U N G   K A N G

H E I   S H U I

C H I A N G

Sung pan Ho

C H O S - C H I A

Shih-ch

D A M B A

S O M O

Mao Chou

Che

Lei-ku-ping

H S U - C H I N G

C H I A R U N G   T R I B E S

Li-fan Ting

An Hsien

H O F A N

B A D I -

B A W A N G

F E U D A L   S T A T E S

Wên-chuan Hsien

Hsiaoshi

W A S S U

Mien-chu Hsien

W O K J I

Sheng-ko-chung

Kuan-cha

Reh-lung-kuan

PAN-LAN SHAN

Ni-yit-kien

NIU TOU SHAN

CHAU LANG SHAN

Shui-mo-kou

Hsüan Kou

Pêng Hsien

Shih-fang Hsien

Tê-y

Pan-ku-chiao

Ta-wei

Hei-shih-chiang

Kuan Hsien

Hsin-fan Hsien

Han Cho

Yo tsa

Mon-kong ling

Chung-ning Hsien

Hsin-tu Hsien

Chin tang

Romi-chango

Tung-ku

Tung-lu-fang

Pi Hsien

M U P I N

Wên-chiang Hsien

CHENGTU

Mao-niu

Ta-yi Hsien

Shung ching Chou

Shuang-liu Hsien

Kuei-yung

Hsin-tien-tzu

Tien-tsuan Chou

Kiung Chou

Hsin-ching Hsien

1

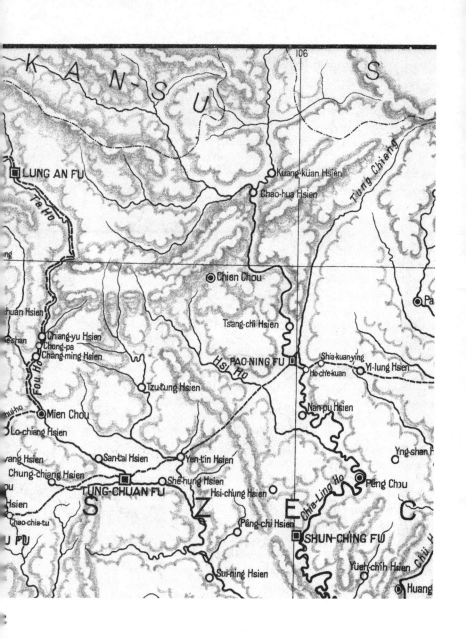

KAN-SU

LUNG AN FU

Ta Ho

106

Kuang-küan Hsien

Chao-hua Hsien

Tung Chiang

Chien Chou

Pa

huan Hsien

e-shan

Chiang-yu Hsien
Chung-pa
Chang-ming Hsien

Tsang-chi Hsien

PAO-NING FU

Shia-kuan-ying

Yi-lung Hsien

Fou Ho

Hsi Ho

Ho-ch'e-kuan

Tzu-tung Hsien

Nan-pu Hsien

huì-ho

Mien Chou

Lo-chiang Hsien

Yng-shan

vang Hsien

San-t'ai Hsien

Yen-t'in Hsien

Chung-chiang Hsien

ou

TUNG-CHUAN FU

She-hung Hsien

Hsi-chung Hsien

Pêng Chou

Chia-Ling Ho

Chao-chia-tu

S

Z

E

C

Hsien

U FU

Pêng-chi Hsien

SHUN-CHING FU

Chü H

Sui-ning Hsien

Yüeh-chih Hsien

Huang

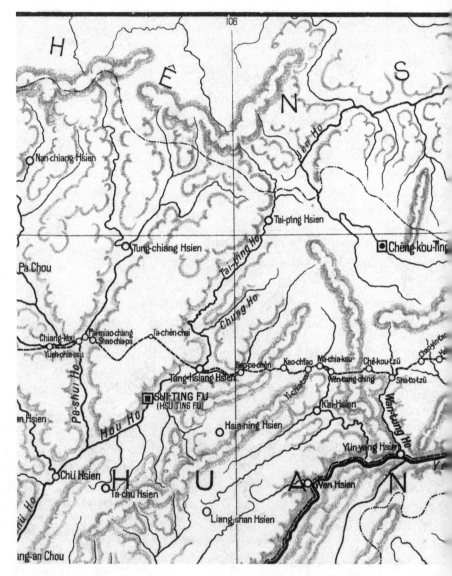

3

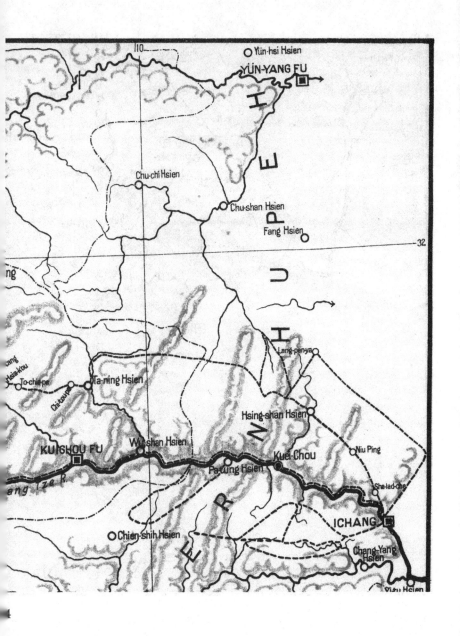

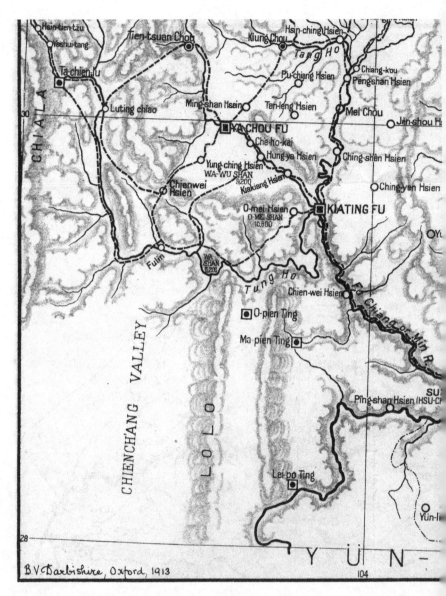

B.V. Darbishire, Oxford, 1913

5

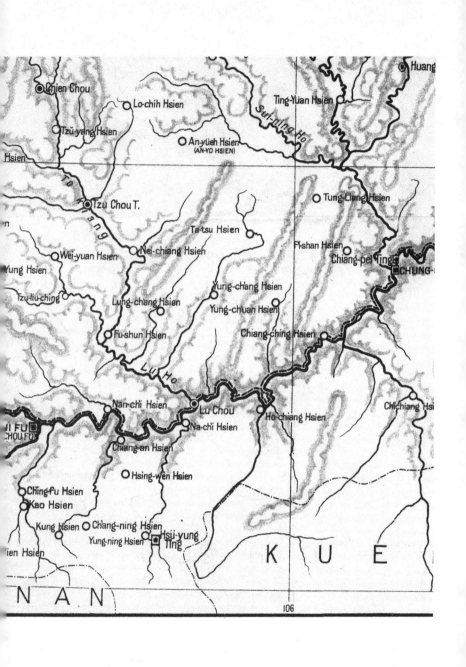

Chien Chou

Lo-chih Hsien

Tzŭ-yang Hsien

Ting-Yüan Hsien

Sui-ning Ho

Huang

An-yüeh Hsien
(AN-YO HSIEN)

Hsien

To Kiang

Tzu Chou T.

Tung-Liang Hsien

Ta-tsu Hsien

Pi-shan Hsien

Wei-yuan Hsien

Nei-chiang Hsien

Chiang-pei Ting

CHUNG-

Yung Hsien

Tzŭ-liu-ching

Yung-chang Hsien

Lung-chang Hsien

Yung-ch'uan Hsien

Fu-shun Hsien

Chiang-ching Hsien

Lu Ho

Nan-chi Hsien

Lu Chou

Ho-chiang Hsien

Chichiang Hsi

Na-chi Hsien

HI FU
CHOU FU

Chiang-an Hsien

Hsing-wen Hsien

Ch'ing-fu Hsien

Kao Hsien

Kung Hsien

Chang-ning Hsien

Hsü-yung Ting

Yung-ning Hsien

ien Hsien

K U E

N A N

106

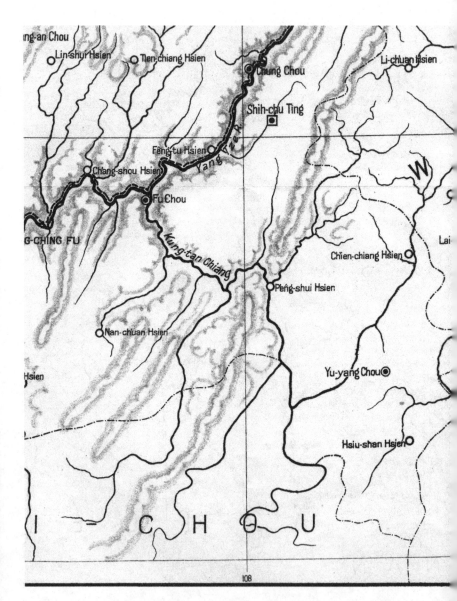

ng-an Chou

Lin-shui Hsien

Tien-chiang Hsien

Chung Chou

Li-chuan Hsien

Shih-chu Ting

Feng-tu Hsien

Yang tze R.

Chang-shou Hsien

Fu Chou

Kung-tan Chiang

G-CHING FU

Lai

Chien-chiang Hsien

W

Peng-shui Hsien

Nan-chuan Hsien

Yu-yang Chou

Hsien

Hsiu-shan Hsien

I - C H O U

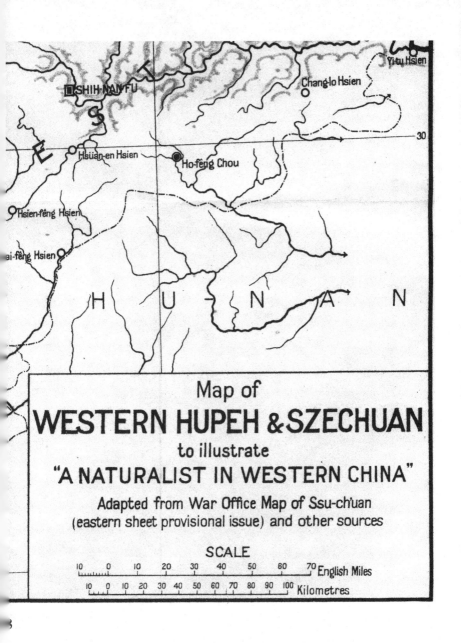

Map of
WESTERN HUPEH & SZECHUAN
to illustrate
"A NATURALIST IN WESTERN CHINA"
Adapted from War Office Map of Ssu-ch'uan
(eastern sheet provisional issue) and other sources

SCALE

INDEX

Abelia chinensis, i. 19
  Engleriana, i. 77
  parvifolia, i. 19, 120
Abies Delavayi, i. 225, 234, 247, 250;
  its wood, ii. 19
  squamata, i. 197, 198
Abutilon Avicennæ, i. 108 ; ii. 81
Acanthopanax aculeatum, i. 109
Acer Davidii, i. 194, 224
  griseum, i. 39, 41, 58.
  oblongum, i. 20
  pictum, var. parviflorum, i. 194
Aconite, medicinal, i. 124
Aconitum Hemsleyanum, ii. 40
  Wilsonii, i. 124 ; ii. 40
Acroglochin chenopodioides, ii. 62
Actinidia chinensis, i. 31, 32, 57,
  173 ; ii. 32.
  Kolomikta, i. 174
  rubricaulis, ii. 32
Acupuncture, ii. 34
Adenophora polymorpha, i. 21
Adiantum Capillus-Veneris, i. 21
  pedatum, i. 134
Adina globiflora, i. 19
  racemosa, i. 74
Æsculus Wilsonii, i. 38, 53, 224, 240
Æthopyga dabryi, i. 46
Agaricus campestris, ii. 63
Agriculture, ii. 48; skill in, 50
Agrimonia Eupatoria, ii. 11
Ai-chiao, ii. 61
Ailanthus glandulosa, i. 20 ; ii. 78
  Vilmoriniana, i. 57, 153 ; ii. 78
Ai Lau-tsze, ii. 147
Ailuropus melanoleucus, ii. 182
Ailurus fulgens styani, ii. 182
Ai Yang-tsze, ii. 153
Albizzia lebbek, i. 88 ; its wood, ii. 22
Aleurites Fordii, i. 17; ii. 5, 64 ;
  distribution of, 65 ; its import-
  ance, 67
  montana, ii. 64, 65
Alder, wood, i. 101
Aldridge, Dr., ii. 150
Allium Cepa, ii. 60
  chinense, ii. 60
  fistulosum, ii. 60
  odorum, ii. 60

Allium sativum, ii. 60
Almond, a new, ii. 27
Alniphyllum Fortunei, i. 232
Alnus cremastogyne, i. 101, 109, 223,
  227, 234 ; ii. 5, 23
Amaranthus paniculatus, ii. 63
Amdo country, ii. 145
Amelanchier asiatica, var. sinica, i.
  32
Amorphophallus konjac, i. 117 ; ii.
  60
Amphicome arguta, i. 189 ; ii. 11
Anaphalis contorta, ii. 62
Anemone japonica, i. 17, 21 ; ii. 10
  vitifolia, i. 127
  vitifolia, var. alba, i. 240
Angelica polymorpha, var. sinensis,
  i. 47, 131 ; ii. 40
An Hsien, city of, i. 116
Animals, game, ii. 144 ; miscell-
  aneous, 190
An-lan chiao, the, i. 171
Antelope, the Thibetan, ii. 160
Antheræa pernyi, ii. 78
Antimony, ii. 198
Apium graveolens, ii. 62
Apple, a fine Crab, i. 56
Apples, ii. 28
Apricot, ii. 26 ; Japanese, 27
Arachis hypogæa, ii. 30, 61
Aralia chinensis, i. 49, 248
  quinquefolia, ii. 37
Archangiopteris, ii. 13
Arctium major, i. 82
Arctous alpinus, var. ruber, i. 136
Areca catechu, ii. 37
Areca-nut, ii. 37
Aristolochia moupinensis, i. 250
Arundinaria nitida, i. 61, 247 ; ii.
  62
Asarum maximum, i. 21
Asbestos, ii. 199
Asmy, Dr., ii. 35
Aspidistra punctata, i. 17
Astilbe Davidii, i. 46, 132
  grandis, i. 46
  rivularis, i. 124, 127
Attacus cynthia, ii. 78
Aucuba japonica, ii. 10

Poppy, a yellow-flowered, i. 56
Poppyworts, i. 138, 181, 199; ii. 9
Populus lasiocarpa, i. 56, 57; fences of, 57
Silvestrii, i. 20, 35
Porphyra vulgaris, ii. 63
Potash salts, i. 117; manufacture of, 125, 247
Potato, Irish, disease of, i. 50; ii. 58
Potentilla anserina, ii. 11
chinensis, i. 21
discolor, i. 21; ii. 63
fruticosa, i. 199; ii. 8
multifida, ii. 63
Veitchii, i. 199, 250; ii. 8
Poterium officinale, ii. 11
Po-ts'ai, ii. 63
Pratt, A. E., 221; ii. 123, 125, 187
Prickly Oak, i. 181
Pride of India, i. 96, 100
Primrose, a blue, i. 38
Chinese, i. 20, 29
Primula Cockburniana, i. 196; ii. 9
involucrata, ii. 11
nivalis, i. 181
obconica, i. 20, 33; ii. 3, 47
ovalifolia, i. 38, 248
Prattii, i. 248
pulverulenta, ii. 9
sibirica, i. 184
sikkimensis, i. 179, 198; ii. 9, 11
sinensis, i. 17, 29; ii. 47
Veitchii, i. 179; ii. 9
vincæflora, i. 179, 181
violodora, i. 39
vittata, ii. 9
Privet, i. 99
Prunus Armeniaca, ii. 26
Davidiana, ii. 26
dehiscens, ii. 27
involucrata, ii. 28
mira, i. 203; ii. 26
mume, ii. 27, 43
Padus, ii. 11
Persica, ii. 25
salicina, ii. 27
serrula, var. tibetica, i. 197
triloba, ii. 43
Przewalski, N. M., ii. 4
Pteridium aquilinum, ii. 63
Pteris longifolia, i. 21
serrulata, i. 21
Pterocarya Delavayi, i. 224
hupehensis, i. 33, 126
stenoptera, i. 20, 223; ii. 5
Pteroceltis Tatarinowii, i. 33
Pterostyrax hispidus, i. 42, 53, 224
Pu-chih-ts'ao, ii. 83
Pucrasia styani, ii. 120

Pucrasia xanthospila, ii. 119
xanthospila, var. ruficollis, ii. 120
Pueraria Thunbergiana, i. 19; ii. 63
P'uêrh tea, ii. 97; uses of, 98
Pulse, varieties of, ii. 55, 56
Punica Granatum, ii. 30
Pun-tsao, or Herbal, ii. 35
P'u-tao-tzu, ii. 30
Pyracantha, crenulata, i. 17, 19; ii. 98
Pyramid, the, i. 15
Pyrola rotundifolia, i. 250
Pyrus sinensis, ii. 28
ussuriensis, ii. 28

Quail, ii. 134
the Bustard, ii. 135
Quercus aliena, i. 58; ii. 21, 78
Fabri, ii. 78
Ilex, var. rufescens, i. 197
serrata, i. 20, 75, 100, 222, 223; ii. 21, 78; cupules, use of, 87
variabilis, i. 58; ii. 21, 78; cupules, use of, 87
fungus culture on, i. 38
Quince, ii. 28

Rafts, bamboo, ii. 140; wild-fowl shooting from, 141
Railway, Hankow-Szechuan, i. 11, 32, 72, 114, 115; ii. 88
Ramie fibre, ii. 82; trade in, 82
Ranunculus acris, ii. 11
repens, ii. 11
sceleratus, ii. 11
Rape, Chinese, ii. 60
Raphanus sativus, ii. 59
Ready, Oliver G., ii. 107
Rebellion, the, i. 114, 115; progress of, ii. 200
Red Basin, i. 3; boundaries of, 64; area, 64; geology, 64, 69
agriculture in, i. 67; crops, 67; fruits, 68; mineral wealth, 69
Reeves, John, ii. 47, 114
Reh-lung-kuan, village of, i. 182
Rehmannia angulata, i. 21, 30, 31
Henryi, i. 35
Reinwardtia trigyna, i. 17, 21
Religious communities, preservation of trees by, ii. 45
Rhamnus davuricus, i. 19; ii. 87
tinctorius, ii. 87
Rhea, ii. 82
Rheum Alexandræ, i. 199
officinale, ii. 39
palmatum, var. tanguticum, i. 131; ii. 38, 39
Rhinopithecus bieti, ii. 191
brelichi, ii. 192
roxellanæ, ii. 191

INDEX

Yilung Hsien, city of, i, 97 ; cotton
in, 97
Yin-chên Mu, ii. 19
Ying Mu, ii. 21
Ying-tao, ii. 27
Yō-tzu, ii. 25
Yuang-ma, ii. 82
Yüan-kou, i. 121
Yuan Shih Kai, i. 115 ; ii. 207
Yuen-fang, village of, i. 92
Yu-la shu, ii. 45
Yün-tou, ii. 56
Yunyang Hsien, salt in, i. 77
Yu-pangtzu, i. 82

Yu-p'o, ii. 41
Yu-shih-tzu, ii. 73
Yu-tsao-chio, ii. 72
Yü-yü-tien, hostel at, i. 178

Zanthoxylum Bungei, ii. 62
Zappey, Walter R., ii. 106, 114, 118,
125, 137, 142, 147, 148, 151, 153,
154, 157, 168, 170, 172, 179, 189,
192
Zea Mays, ii. 53
Zingiber officinale, ii. 59
Zizania latifolia, ii. 63
Zizyphus vulgaris, ii. 30

Printed by
MORRISON & GIBB LIMITED
Edinburgh

Printed in the United States
By Bookmasters